Naked

Agency

A THEORY IN FORMS BOOK   *Series Editors*   Nancy Rose Hunt and Achille Mbembe

Naked

Agency   Genital Cursing and Biopolitics in Africa

NAMINATA DIABATE

Duke University Press *Durham and London* 2020

Designed by Courtney Leigh Baker
Typeset in Garamond Premier Pro and Quadraat Pro
by Copperline Book Services

Library of Congress Cataloging-in-Publication Data
Names: Diabate, Naminata, [date] author.
Title: Naked agency : genital cursing and biopolitics in Africa /
Naminata Diabate.
Description: Durham : Duke University Press, 2020. | Series: Theory
in forms | Includes bibliographical references and index. Identifiers:
LCCN 2019013469 (print) | LCCN 2019016277 (ebook) ISBN
9781478007579 (ebook)
ISBN 9781478006152 (hardcover : alk. paper)
ISBN 9781478006886 (pbk. : alk. paper)
Subjects: LCSH: Nudity—Political aspects—Africa, Sub-Saharan. |
Women—Political activity—Africa, Sub-Saharan. | Human body—
Political aspects—Africa, Sub-Saharan. | Political activists—Africa,
Sub-Saharan. | Social action—Africa, Sub-Saharan. Classification:
LCC BJ1500.N8 (ebook) |
LCC BJ1500.N8 D533 2020 (print) | DDC 391/.20967—dc23
LC record available at https://lccn.loc.gov/2019013469

Cover art: Adapted from Opération Kodjo Rouge I, July 24, 2011.
Source: Civox.net.

Duke University Press gratefully acknowledges the support
of the Hull Memorial Fund of Cornell University, which provided
funds toward the publication of this book.

Dedicated to my
mom,
Sitta Camara,
and my dad,
Morikounady Diabate,
for their faith in me, a girl.

CONTENTS

ACKNOWLEDGMENTS

Even the sheer mention of genital cursing elicited fear and concern in my childhood neighborhood of Abidjan in Côte d'Ivoire. Over the years of writing this book, my major driving force turned into a challenge: How do I write on a subject that is both personal and intellectual? In seeking to understand the gesture, I benefited from the intellectual generosity and friendship of countless persons and institutions. In the initial stages of this project, Lisa L. Moore's robust questions and challenges gave me the boost that I needed to devote myself fully to the book. Ever since, Lisa has provided academic rigor and spiritual guidance.

I am honored that Neville W. Hoad agreed to work with me. Shared with humor, his direct questions and incisive comments pushed me toward stimulating theoretical terrains. At the University of Texas at Austin, various professors and mentors allowed space for intellectual freedom and professional development: Elizabeth Richmond-Garza, the late Barbara Harlow, Jennifer Wilks, Helene Tissières, and Matt Richardson. Over the years, Professor John S. Butler has been a continuous source of inspiration with his humbling principle, "three articles under review at all times."

In my journey, I encountered the collegial and inspiring Department of Comparative Literature at Cornell, which awarded me the 2012–14 Andrew W. Mellon Postdoctoral Diversity Fellowship. Under Gerard Aching's distinguished leadership, I benefited tremendously from the knowledge and professionalism of outstanding Cornell faculty members who participated in the two years associated with the seminar. The focal themes, "The Native Subject" and "Divergent (Post)Modernisms," invited enriching and sometimes heated debates. Cornell faculty participants included Leslie Adelson, Judith Byfield, Eric Cheyfitz, Judith Peraino, Suman Seth, Arnika Fuhrmann, Margo Crawford, Deborah Starr, Sarah Murray, Magnus Fiskesjö, and Richard Miller. Their mentoring advice and attentive comments on parts of the book opened my eyes to additional theoretical frameworks.

I wish to thank my "diverse" post/doctoral fellows, as we called ourselves, Christopher Pexa, Lorenzo Perillo, Murad Idriss, and Nicole Giannella. Our on- and off-campus gatherings made my first years in Ithaca thrilling. My often-challenging conversations on a few conceptual choices further fueled my determination to make my arguments as clear as possible for a wider readership.

Throughout my first years as an assistant professor, I have been lucky to benefit from the intellectual generosity of many colleagues in the department and beyond. The fruitful intellectual conversations with Debra Castillo, Natalie Melas, Tracy McNulty, Calum Carmichael, Andrea Bachner, Thomas McEnaney, Cathy Caruth, and Antoine Traisnel demonstrate the power of a healthy scholarly community. My department mentor Debra Castillo deserves special mention. She read each chapter and more with unmatched rigor and enthusiasm. For their unmatched professionalism regarding logistical support that saved me headaches, I thank Sue Besemer, Dorothy Vanderbilt, and Laura Kipfer.

The longer it took to complete this book, the more debt I incurred and the more hats (colleague, interlocutor, mentor, ally, and friend) some of those mentioned here had to put on. I wish I had the material space to acknowledge all their different and complementary contributions.

Members of my writing group—Lucinda Ramberg, Kate McCullough, Saida Hodžić, and Arnika Furhmann—had to wear multiple hats. They commented on various chapters over body- and mind-nourishing dinners. Our collegiality and intellectual generosity meant much more to me than I can express here. I only hope this book might distantly reflect their exceptional professional and intellectual achievements.

I am grateful to many organizations and programs whose contributions made this book possible. I thank the 2013 Cornell Summer Institute for Critical Theory and our seminar leader, Ian Baucom. The institute not only allowed me to meet friends and intellectual interlocutors but also stood out as one of the most challenging short periods in my intellectual maturation. Under the guidance of Yael Levitt, the Cornell Office of Faculty Development made a useful impact on this book through funding the manuscript workshop, which Gerard Aching organized in spring 2017. At the workshop, E. Frances White, Dorothy Hodgson, Judith Byfield, and Debra Castillo made the manuscript a more readable and nuanced set of reflections.

A faculty fellowship (2016–17) at the Cornell Society for the Humanities under the rigorous stewardship of Timothy Murray allowed me to devote my time to completing this book. Tracy McNulty, Debra Castillo, and Karen Pinkus were more than colleagues; they were allies. And so were Wendy Laura Belcher of Princeton University and Shalini Puri of the University of Pittsburgh. At the society, my "skin fellows," as we called ourselves, invited me to elaborate and lay out clearly my arguments for a wider readership: Andrea Bachner, Alicia Imperiale, Samantha N. Sheppard, Stacey Langwick, Pamela Gilbert, Nancy Worman, Karmen MacKendrick, Gemma Angel, Gloria Chan-Sook Kim, Ricardo A. Wilson II, Seçil Yilmaz, Kevin Ohi, Elyse Semerdjian, Erik Born, Daniel Smyth, Emily Katherine Rials, Alana Staiti. Dinners, karaoke, and laughter with some of them made completing the book a much more enjoyable experience. And, in some, I have made lifelong friends.

My heartfelt thanks go to colleagues at Cornell: Judith Byfield, Dagmawi Woubshet, C. Riley Snorton, Siba Grovogui, Carole Boyce Davies, Marie-Pat Brady, and Salah Hassan. Olufemi Taiwo deserves special gratitude for encouraging me to see my cultural background as complex and deserving of rigorous intellectual engagement.

Thanks to Elizabeth Ault at Duke University Press, whose belief in this project, and insistence that this reluctant writer undertake the necessary changes, in retrospect shaped this book for the better. With grace and yet firmness, she kept this project moving forward. Thank you to Kate Herman for helping me figure out permissions, maps, illustrations, and such. I would have been lost in this process without her professionalism. I am grateful to Nancy Hunt and Achille Mbembe, the series editors, for their theoretical rigor. I am also grateful to the readers of the manuscript for their serious engagement with this book and their robust conceptual and structural comments.

I thank Amanda Masterson, Stephanie Jones, and Sheri Englund for their editorial assistance with various aspects and over various stages of this book. Amanda has read this book more times than anybody else has, and for that, I also thank her.

Various families from around the world have nourished my various selves while I was writing this book. First of all, my siblings, here or gone, have inspired my journey toward selflessness. I thank Mamadou, Bakary, Vamara, Sene, Fanta, Adjara, Yacouba, Yaya, Mbene, and Chaka for allowing me to be their sister. Thanks to my sister-friend Somy Kim and my brother from another continent, Assem Nasr, for their nourishing friendships. Thank you to their spouses and children, Anthony Melvin (Marwan and Mirae) and Sonia Checchia (Carma and Nellie), for always keeping their doors open for me. My peers in graduate school—Michal Raizen, Lanie Millar, Tessa Farmer, and Joanna Sellman—are more than friends; they have become true intellectual interlocutors.

My long and passionate conversations on various forms of resistance during moments of political instability and social upheaval in Côte d'Ivoire convinced me that working on female defiant disrobing is more than an academic exercise.

The subject carries material consequences for some of us. I am grateful to these members of my African family in the United States—Moussa Silue, Mamadou Kane, Kapo Martin Coulibaly, Gnimbin Ouattara, Salimata Coulibaly, and Alexie Tcheuyap.

My Austin family—Master Tim Bugh, John R. Kelly, Nailah I. Akinyemi-Sankofa, and Shelley Damon—introduced me to Texas, its warmth, and its Tex-Mex cuisine, all things I have found comforting in the United States. My gorges Ithaca family—Karen Williams, Dr. Jennifer Orleans, Marie-Helene Koffi-Tessio (who read parts of this book), Aaron Pichel, Esq., Gail Dennis, and Stephanie Tchuente—have provided me much-needed emotional and intellectual support. They all made Austin and Ithaca homes away from home.

Exceptional Nakedness

It is a foolish woman who degrades womanhood. —**Malinké proverb**

This was Egbiki, the secret, nocturnal ritual of female genital power. It was an act of spiritual warfare, a critical and dangerous enterprise that the women were undertaking on behalf of the whole village. —**Laura Grillo**, *An Intimate Rebuke*

In July 2002, hundreds of female protesters in Nigeria occupied Chevron Texaco's export terminals and properties, holding hostage seven hundred American, British, Canadian, and Nigerian workers by threatening to get naked. The women demanded employment for their families and basic infrastructure, such as schools, motorable roads, and water and electrical systems. Major international news channels—including ABC News, BBC, Associated Press, and Inter Press Service—brought the incident to the world's attention. Eventually, Chevron Texaco agreed to negotiate with the protesters, leading Associated Press journalist D'Arcy Doran to claim, "For the women, what started out as an act of desperation became a method to victory" (2002). Asked about their abilities to take men hostage without guns or heavy artillery, Helen Odeworitse, a spokesperson for the women in this extraordinary seven-day protest, explained, "Our weapon is our nakedness" (BBC News 2002).

Scholars from Africa and elsewhere have written insightfully on the global significance of the 2002 event to later protest movements. Some highlight the women's indisputable power by drawing on ethnographic data. Others attribute the recent upsurge of resistant nakedness internationally to the Nigerian women's threat to strip naked to shame multinational oil companies.[1] On her website, baringwitness.org, American activist Donna Sheehan (2002a) identified the women's action as the inspiration to protest in California against the war in Iraq. In academia, most scholars use the terms *genital power* and *genital cursing* for their arguments about the Niger Delta event, independent of whether that was the women protesters' name for their actions.

In this book, I offer a different reading of the July event by considering the women's feelings and the reactions of their targets, supporters, bystanders, and witnesses. For instance, Odeworitse's statement "Our weapon is our nakedness," which was issued after what appears to have been a victory, or at the least a "deal," has become a sound bite in most news reports of defiant self-exposure in Africa. Speaking of the victory, one of the protest leaders, Anunu Uwawah, announced that the oil company pledged to hire more villagers and build amenities (BBC News 2002). Although powerful, Odeworitse's declaration obfuscates several aspects of the event: the constitution of the women's collective, the kind of organizing behind the forceful occupation, the women's expectations, the reactions of their targets, the way the declaration was distributed among media and the possible uses to which it was put, and more. In a world where the idea of Africa as an ontological oddity persists, the self-aggrandizing declaration, made after the "deal," and its endless appearances in Western media participate in portraying the continent and its practices as fundamentally singular. A newspaper report naturally truncates, condenses, and simplifies complex statements, especially for Western readers. Were the women hoping for their "nakedness" to be the most significant aspect of their protest? Perhaps the women did not intend or foresee the uses and misuses of their strategy. By calling it *nakedness*, is Odeworitse stripping the act of its complex connotations to make it legible to the international reader? In her statement, how inclusive is the word *our*? Did the *deal* reflect the women's expectations? Were all of the mature women ready to disrobe to curse, and what was the act to be called?

Unless the women specifically call their disrobing *genital cursing*, as many scholars have done, it is misleading to call it that.[2] The July 2002 standoff is not just a definite event with knowable and foreseeable implications. It is, rather, a complex event interwoven with narratives of desperation, agency, satisfaction, and exploitation. Broadly, because of the attention that nakedness in protest begets, and because of its nature as signifying shorthand, the gesture is subject to mediation and translation, and it demands exploration. As the most universal and yet the most highly context-driven mode of dissent, insurgent nakedness is not just one thing with multiple interpretations. It is many things. It is a different code to decipher deeper cultural and societal accounts each time it is used, not only in its interpretation but also in its constitution. These are the questions and readings that I emphasize in my exploration of these dramas of desperation, anger, agency, and victory. I suggest that the event and similar others should be read on their own terms but also in conjunction with larger sociopolitical and aesthetic frameworks.

Naked agency is the term that emerges from my innovative readings and that I use to designate the dynamic cycle of power and vulnerability that involves the women and their targets. From the event and its manifestations emerges a cycle of contestation, exploitation, and misreading that may differ from the women's original plan. In this dynamic, the agency of the women, their targets, and other stakeholders are simultaneously co-constitutive, instrumentalized, precarious, and triumphant. *Naked agency* is both a concept and a reading praxis that privileges the dialectical movement between these fluctuating narratives for a more comprehensive understanding of resistant self-exposure. To complement *naked agency*, I interchangeably use more neutral terms such as *defiant disrobing, naked self-exposure, assaultive nakedness,* and *intentional nudity*. These terms contrast with *genital cursing, genital power,* and *female genital power* (FGP). Later, when relevant, I use indigenous words for these acts.

A Tripartite Approach

My multidisciplinary approach and insights into naked agency are revealed through the exploration of dozens of protest demonstrations throughout both Francophone and Anglophone Africa and through attention to multiple media, including documentaries, social-media material, literary fiction, and narrative film. In relation to the women of the Niger Delta's threat to strip naked, two internationally circulating visual renderings speak to the long history, and the importance, of the threat. In 1994 Bruce Onobrakpeya created his plastographic piece *Nudes and Protest* to stage how elderly women of the Niger Delta express their grievances against unrestrained rule and mismanagement of natural resources.[3]

In the blue-and-green artwork, three rows of about thirty mature naked women face targets that are invisible to the viewer, who holds a dominant position over the standing women because of the high viewing angle. Onobrakpeya's staging is not one of outright confrontation between the women and their targets. Here, the emphasis is on the women because the targets are absent from the frame. Surrounded by what appears to be a bush or shrubs and holding leafy branches as maternal symbols, most of the figures seem to cover their breasts with branches or with their hands. Their facial expressions vary, with some women expressing anger and others looking distracted. Unlike photographs of recent insurgent disrobings with women displaying gestures of defiance, the artwork already invites us to be more nuanced in our engagement with naked protest.

A feature documentary by Candace Schermerhorn, *The Naked Option: A Last Resort* (2010), also documents another group of women's expectations and their rationales for threatening to perform the naked ritual in protest against the actions of oil companies in 2002. As the women discovered during negotiations, their initial demands clashed with the reality of multinational oil companies and their local allies. In this book, I read news reports of the July 2002 event in combination with Onobrakpeya's artwork and Schermerhorn's documentary. Both of these creative works convey something about the women's feelings, the proliferation of insurgent nakedness in Africa, and the reactions that the gesture encounters.

Thus, I resituate the analysis of naked self-exposure within a different set of scholarly conversations—those about our biopolitical era and about insurgent embodiment in contemporary Africa—by choosing a tripartite approach that is descriptive, interpretive, and theoretical. Contrary to the customary postcolonial readings with their emphasis on African cosmogonies versus the universalist, read Eurocentric, interpretations of biopolitics that I explore within this text, my work draws on both African cosmogonies and theories of agency and biopolitics to uncover how certain forms of political participation become paradoxically both legible and illegible in our era. I ask the book's central questions: What kinds of sociopolitical climates and arrangements call for the deployment of self-disrobing in anger? How is the women's agency to be understood and conceptualized in light of the reactions of their targets and of other stakeholders? What is being lost by focusing on precolonial and colonial explanatory frameworks?

Using news reports and historical documents, I describe dozens of protests from the colonial to the postcolonial era (nine decades) in more than twenty Francophone and Anglophone African countries south of the Sahara (see map 1.1 for an overview of protests in the period 1922–2017). Within exceptional and biopolitical conditions, when all else has failed, and when their bodies seem to be all that they have left, the women's insurrectionary gesture often is their last resort. Whether used as an expression of vulnerability, or as a mode of conflict management or resistance, the potency or the inefficacy of their disrobing often stems both from authorizing social structures such as the privileges of motherhood and social cohesion and from prevailing notions of gender and sexuality. Paradoxically, however, their gesture is either exploited or repressed, depending

Morocco

Tunisia

Western
Sahara

Algeria

Libya

Egypt

Mauritania

Niger

Mali

Eritrea

Burkina
Faso

Chad

Sudan

Senegal

Djibouti

Gambia

Guinea-
Bissau

Guinea

Nigeria

Sierra
Leone

Liberia

South
Sudan

Uganda

Ethiopia

Central
African Rep.

Côte d'Ivoire

Ghana

Togo

Benin

Cameroon

Eq.
Guinea

Somalia

Kenya

**Democratic
Republic of
the Congo**

Rwanda

Gabon

Burundi

Congo

Tanzania

Mozambique

Angola

Zambia

Madagascar

Zimbabwe

Malawi

Namibia

Botswana

Swaziland

**South
Africa**

Lesotho

Protests by Country

- 1–2
- 3
- 4–6
- 7–11

Protests by Decade

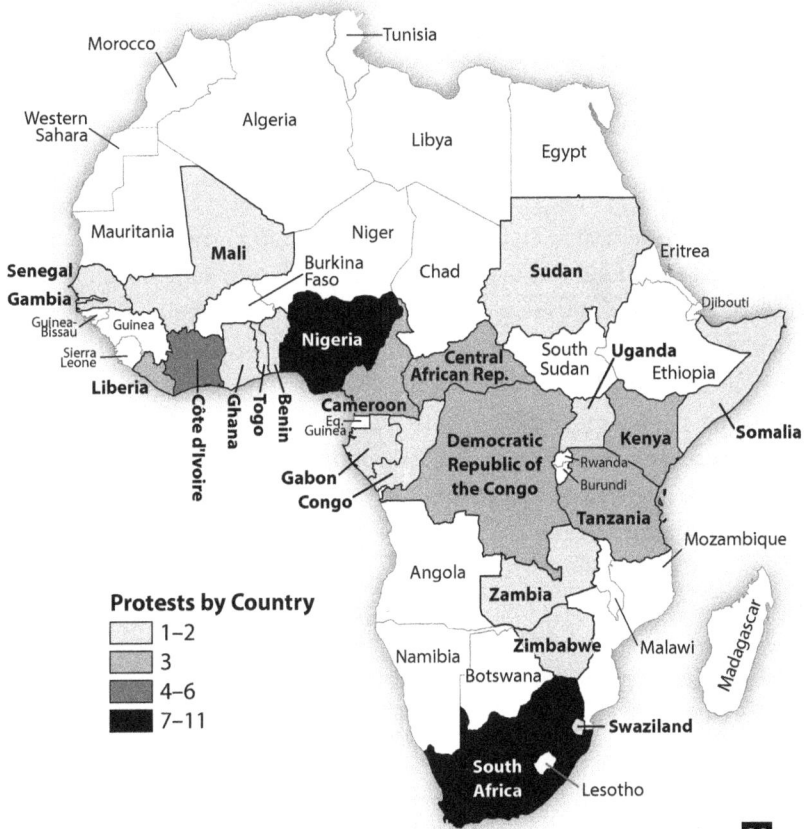

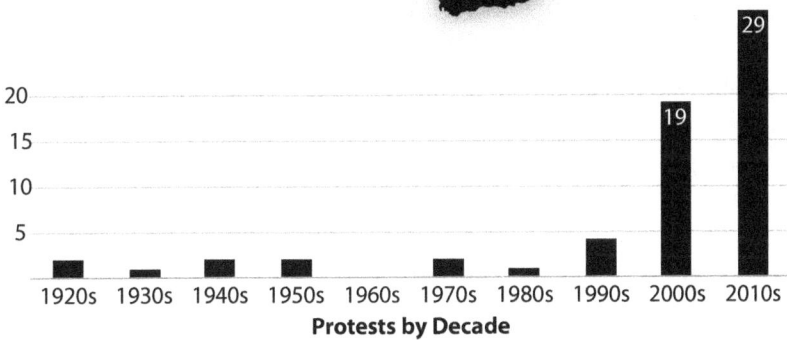

MAP I.I Map of Africa's naked protests (1922–2017). Courtesy of Tim Stallmann.

on the disintegration of these authorizing structures within so-called political secularism.

Comparing West Africa and South Africa in this study, I demonstrate that pairing the continent of Africa and the word *country* gives the wrong impression. In fact, Africa is a heterogeneous continent with differing nation-states, each with its own political history. I argue that the sense of genital cursing as a means of killing offenders by enlisting ancestors, which has been used as a tactic in events in West Africa, differs from the cross-ethnic use of naked shaming in movements such as that of "Fees Must Fall" by students in South Africa.

On the theoretical level, I examine the contributions that attention to resistant self-exhibition can make in our assessment of political agency when both desperation and a sense of power are operative. Bringing together these assumed opposed states through the lens of biopolitics and postcolonial discussions of agency, I analyze emancipatory political subjectivities that are underestimated in dominant Euro-American reflections but overvalued in African studies. As a defining concept, "biopolitics" makes sense because our era is mainly characterized by what some consider to be the futility of agency and resistance against tyrannical and rhizomatic regimes of power. Moreover, such regimes are mainly identified as producing literal or figurative death (Agamben 1998).

In interpreting the women's insurgent disrobing via the theory of agency, according to which there is no agency outside of constraints, the concept of naked agency expands Giorgio Agamben's (1998) conceptualization of "naked life" so as to contribute significantly to an affirmative understanding of biopolitics.[4] The openness and the nakedness of positions of emancipation and constraint reflect more accurately the unstable political subjectivity that has developed within current sociopolitical arrangements, in which the state of exception (Agamben 2005) and the widespread sense of perishability are now normalized. I propose that naked lives and exposed bodies can still be constituted to defy sovereign power, if only briefly and imperfectly. Although Agamben's concept of *la nuda vita* in *Homo sacer: Il potere sovrano e la nuda vita* (1995) has been popularly translated into English as "bare life," I refer to translations wherein he and others suggest "naked life" rather than "bare life" as being more appropriate. For example, in Agamben's ruminations on nudity in *Nudities* (2011), he relates the figure of *Homo sacer* to naked rather than bare life.[5] Karen Pinkus and Michael Hardt also translate *nuda vita* as "naked life" in Agamben's *Language and Death* (1991), as do Vincenzo Binetti and Cesare Casarino in Agamben's *Means without End: Notes on Politics* (2000). The reference to the figure of *Homo sacer* as "naked life" as opposed to "bare life" carries conceptual implications in Agamben's work. Perhaps "bare," unlike "naked," suggests milder degrees of exposure, when in fact it is the confrontational, shame-inducing, or death-producing aspects of rebellious disrobing, for either the protesters or their targets, that is at the core of the term *naked agency.*

Finally, analyzing more aestheticized genres through an interpretive lens, I highlight the multiple reactions that the women's gesture of disrobing prompts; the women's internal conflicts (whenever knowledge of these is available); and the counterproductive responses of their targets, bystanders, and other stakeholders. Given that the production, experience, and enactment of emotions are central to the performance of aggressive self-exhibition, and given the attention-grabbing nature of most news reports and historical documents with their etiolated accounts of insurgent disrobing, the different and emerging artistic responses that disrobing has elicited from novelists, filmmakers, and other visual artists become crucial. In this book, through an "open," read cross-genre and cross-media engagement, I provide evidence from documentaries, novels, autobiographies, social-media posts, pictorial arts, and narrative films to provide emotion-focused perspectives not yet collected elsewhere. Some of the autobiographies are Gbowee 2011; Maathai 2006; and Wainaina 2011.

In one of my chapters on literary fiction (scene 8), for instance, I use T. Obinkaram Echewa's historical novel *I Saw the Sky Catch Fire* ([1992] 1993) to draw attention to the counterproductive effects of defiant disrobing. Through specific narrative techniques, the novel restages gendered colonial trauma to expose the murderous reactions of the women's targets. My chapters on visual materials (scenes 2 and 5) trace particular contemporary configurations in Africa in which the women are both pawns and political agents. The women are compelled to protest against social and political inequities, but they have limited influence on how their lives will be governed. Specifically, the South African short documentary *Uku Hamba 'Ze (To Walk Naked)* (Maingard, Meintjes, and Thompson 1995) demonstrates constraints on women's autonomy, while Jean-Pierre Bekolo's postmodern film *Les Saignantes* (2005) and Candace Schermerhorn's *The Naked Option: A Last Resort* (2010) both consider various forms of co-optation to which the gesture of disrobing is subject. Such documentaries and novels depict the clearly defined political stakes of defiant disrobing, enabling access to the thoughts and feelings of women who bare their genitalia. In complementing and adding nuance to news reports and their schematic structures, discursive and representational accounts provide other perspectives to the terms of current debates on naked agency.

I also engage social-media material on an interpretive level to argue that the deployment of "genital cursing" fails in certain online spaces. Coverage through

YouTube videos and Facebook posts ends up trivializing what is supposedly the women's most powerful gesture.

The Era of Exposure?

Since the late 1990s, insurgent nakedness has proliferated worldwide. I have documented almost five hundred collective instances of public nakedness around the world in response to many issues. Globalization, capitalism, HIV/AIDS, war, power abuses, Black Lives Matter, animal rights, land disputes, lactivism, environmentalism, and the rise of Donald Trump are all issues that have evoked naked agency. The proliferation of the act shows that it is less the last resort of oppressed, minoritized groups than it is a global response. From Colombia to Mexico, from France to India, from China to Kenya, various sociopolitical male and female actors have publicly exposed their genitals and breasts in political protest. For instance, in 2004 in Manipur, a small state in the eastern part of India that is under heavy military presence, mothers marched without clothes to express vulnerability and outrage about rape. In the United States, the People for the Ethical Treatment of Animals (PETA) campaign "I'd Rather Go Naked Than Wear Fur" (2006–19) or La Tigresa in her "Striptease for the Trees" (2000) actions in California have utilized resistant self-exhibition to prosecute conflicts with opponents.

Protests against Donald Trump during and after his road to the White House were more "naked" than those of any previous protest campaigns (including those in 1968 against the Vietnam War and in 2000 against the Iraq War), with numerous protests and scenes of intentional exposure and self-exposure regarding Trump.[6] For example, covers of the July 2015, October 2016, and March 2018 issues of the *New Yorker* depicted a naked Trump, making him the first American president to have had his body exposed to such an extent. The election year of 2016 also saw naked statues of both Trump and rival Hillary Clinton illicitly erected in Manhattan, as well as the online media publication of seminude photographs of Melania Trump from decades earlier (I. Vincent 2016). Whether it was the depicted nakedness of candidates or spouses, or the assaultive nakedness of protesters and celebrants in the United States or abroad, public disrobing (deliberate or forced) was a major phenomenon in 2016. Scenes of nakedness proliferated in roughly seven genres: nude installations, photography, film, painting, sculpture, comedic viral video, and news reports. A few of these scenes merit attention for the insights that they enable regarding naked protest in our globalized world.

In July 2016, renowned American photographer Spencer Tunick—whose large-scale nude installations draw thousands of naked participants from around the world—organized an installation titled *Everything She Says Means Everything* with one hundred women, just hours before Donald Trump took the stage at the Republican National Convention in Cleveland, Ohio. The naked women held large mirror discs to reflect "the knowledge and wisdom of progressive women" (*Business Insider* 2016). According to Tunick, Trump was a "loser" (*Business Insider* 2016). Other instances of assaultive nakedness were enacted by several other groups of protesters. In early November 2016, for instance, two female protesters stormed Manhattan polling stations with "Trump, grab your balls" written in black paint across one of the protesters' bare chest and stomach. The other woman had written "Hate out of my polls" around her exposed breasts (Moye 2016). The women were reportedly members of the U.S. chapter of Femen, the international radical feminist activist organization that has staged hundreds of topless protests around the world since 2008. In actions in France, Canada, Tunisia, Russia, the United States, and Spain, Femen has often called its topless protests "titslamism," "sextremism," and "topless jihad." After the Femen protesters of November were arrested and released, one of the women who bared her breasts in protest against Trump was quoted as saying, "We use our bodies to express our ideas" (Chan 2016).

Since the 1990s and in more than two dozen Francophone and Anglophone sub-Saharan African countries—including Nigeria, Uganda, Kenya, Togo, Tanzania, Côte d'Ivoire, Gabon, Cameroon, South Africa, Mali, Congo-Kinshasa, Central African Republic, and Ghana—dissident self-exhibition has been appearing with remarkable frequency. Thousands of women in Africa have taken to the streets naked or have flashed their genitals to resist during acute national social and political crises and to punish elected officials.[7] They have also stripped to signify vulnerability, expose distress, or to enlist the support of national or international communities for their causes. Women of Ekiti State in Nigeria who sought to delegitimize their leaders after local electoral rigging in 2008 drew on prevailing shared beliefs about the sanctity of motherhood and social cohesion (K. Ajayi 2010). In 2002 several Ugandan young and middle-aged female opposition activists, wearing only their bras, stormed the Kampala police headquarters and taunted the police, trying to upset their targets (ChimpReports 2012). To achieve these goals, the women took advantage of the social capital of sexuality and of the wider circulation of news and images afforded by the internet and other media.

Uncivil self-exposure has been mobilized in urbanized centers in Africa by women of all ages, educational levels, and occupations.[8] They are widows, civil servants, mothers, street vendors, university students, young, and elderly. The proliferation of these instances reveals the dangerous social and political climates that have compelled some women to use what has historically been a last resort. Hostile counterattacks against protesters—such as legal threats, arrests, verbal attacks, physical brutality, and even murder—further reflect the severity of conditions provoking the gesture. Indeed, naked agency as protest occurs in such diverse contexts as (perceived) electoral injustice, land grabbing, corporate greed, police brutality, the importation of foreign toxic waste, unethical clinical trials conducted by pharmaceutical companies, and adverse neoliberal policies. These circumstances do not define the continent as an ontological oddity. Rather, it is because Africa is a space of incipience, fabrication, and test-driving of stylistics of power that its populations live and die in these conditions.

As in Africa, protesters in the United States strip naked to signal powerlessness and to demonstrate wisdom and strength. However, in the United States, nakedness is also often mobilized to highlight freedom and to celebrate human bodies. A participant in Tunick's nude installation said, "Being nude in public is a freeing kind of experience. We want to be able to show our bodies and be able to express to people who are telling us what to do with our bodies that they have absolutely, positively no right to tell me or any of those other women there what to do with our bodies" (Bigley 2016). In this statement, contestation, play, and exhibitionism coalesce to expose arbitrary rule. In Africa, defiant disrobing is overwhelmingly performed in a solemn mood, while in the United States, the festive mood of most of the protest demonstrations often blurs the lines between celebration and vulnerability. Perhaps, for these American protesters, celebration and the expression of vulnerability are not antithetical, as entertainment is often a feature of serious matters, for example, some legal trials in the United States. The combination has invited, however, charges of social deviance and has likened those who protest in the nude to those who engage in mooning, streaking, flashing, or exhibitionism. Unsurprisingly, these criticisms have dulled the political efficacy of this kind of self-exposure in the West.[9]

What accounts for this increasing mobilization of uncivil nakedness in political and social contention? In 2007 Jean-Luc Nancy, a major thinker and writer about literal and metaphorical nakedness, justified the proliferation as an index of the nature of our era: the era of exposure. Ours is an age that "feels particularly exposed, that is particularly stripped of ideological vestments" (quoted in Nicklaus 2007).[10] This argument, which is time- and place-specific, is informed by the Fall in the Garden of Eden, a fundamental feature of the dominant Christian accounts of nakedness as weakness, lack, and shame for the unclothed person (Agamben 2011; Derrida 2002; Ferrari and Nancy 2014; Grosz 2006; Nancy 1993). This version of nakedness, which has tended to dominate the global conceptual landscape, takes the North Atlantic world as the source of empirical data for reflections that may be uncritically perceived as universal.

A practice documented since medieval times in Africa (Diabate 2016), the aggressive disrobing of mature women continues today to provoke intense debates in newspapers, pictorial arts, oral tradition, narrative film, documentaries, novels, and autobiographies, as well as on social media (e.g., YouTube, Facebook, listservs, and personal blogs). The publicness, the cultural potency, and the extreme sociopolitical conditions required to mobilize protesters explain the interest that resistant self-exposure has received from scholars of many disciplines, including anthropology, psychiatry, political science, sociology, and history.[11] These scholars explore defiant disrobing to examine gender relations and power differentials in sub-Saharan African societies, particularly those marked predominantly by patriarchal regimes.[12] Their studies draw on deep cultural meanings, what I call the "cosmological," to explain the potency of women's power. For instance, in *An Intimate Rebuke: Female Genital Power in Ritual and Politics in West Africa*, a recent and insightful cultural anthropological study of female genital power (FGP) among the Abidji of Côte d'Ivoire, Laura Grillo (2018) establishes the potency of women's gesture in West African societies. She describes one society's activity thus: "This was Egbiki, the secret, nocturnal ritual of female genital power. It was an act of spiritual warfare, a critical and dangerous enterprise that the women were undertaking on behalf of the whole village" (1). Like Grillo with her term *FGP*, other scholars use terms such as *genital power* (Stevens 2006), *genital curse* (Bastian 2005), *sexual insult* (Ardener 1973), and *bottom power* (Day 2008). The collective nature of the women's act has been framed as *shaming parties* (Edgerton and Conant 1964), *women's mobbing* (P. Spencer 1988), and *punitive delegation* (Talle 2007).[13] The cosmological interpretation by both scholars and politicians has been mobilized to highlight women's absolute power effects.

Unlike these studies and their disciplinary and geographical foci, the claims of *Naked Agency* are larger in scope, covering sub-Saharan Africa. Moreover, the book's specificities lie in its theoretical orientation, wide geographic scope, and attention to multiple aestheticized genres, including literature, film, artwork, and social-media posts. For instance, *Naked Agency* and *An Intimate Rebuke* by Laura Grillo move forward the conversation on defiant disrobing in Africa by engaging complementary but different disciplinary and methodological frameworks. In light of these disciplinary specificities, Grillo and I reach different conclusions, particularly how to understand the proliferation of naked protest on the continent. Whereas *Naked Agency* highlights the unstable nature of positions of power by paying attention to women's targets and other stakeholders,

An Intimate Rebuke focuses on the women and their worldviews within the context of a constantly changing world.

The Cosmological

At the core of this interpretative framework are customary beliefs, many of which may no longer be operative. Among the Igbo of Nigeria, for instance, some of these beliefs include the "potency of women's fertility, fear of senior women's sexuality, and prohibitions against incest, particularly among members of uterine households" (Bastian 2005, 46).[14] Incest is committed in these circumstances when males in the community look at elderly women's exposed genitalia.[15] Among the Yoruba of Nigeria, "traditionally, the onlookers are also guilty by association. Customarily, they were expected to also go naked in solidarity with the women's cause" (Oyeniyi 2015, 153). Where social cohesion requires the cooperation of the community in upholding its core values and retributive schema, people fear being associated with such a crime and its resultant social stigma.

The potency of women's bodies is said to lie in the fact that many societies—including the Akan of Ghana and Côte d'Ivoire, the Beti of Cameroon, the Malinké of Mali, the Igbo of Nigeria, and the Kikuyu of Kenya—consider women's bodies as both life giving and life taking. Ritual nakedness, performed by members of female institutions that are often known, problematically, as secret societies, is part of indigenous religious beliefs, according to which both maleficent and beneficent spirits reside in women's bodies. (Although I consider the term *secret society* problematic because it was used derogatively in the colonialist literature, I will continue to use it in the book for lack of a better term.) Their life-giving ability, "the ultimate in human power" (Brett-Smith 1995, 33), explains the fear that the gesture provokes in men, even in armed men. This potency is called *evu* in Beti (Ombolo 1990) and *nyama* (the principle of evil) in Malinké and Bamana (Brett-Smith 1995; Even 1939; Koné 2004).[16]

It is this putative power that is enshrined in the Malinké proverb, "It is a foolish woman who degrades womanhood." Among the Bamana of Mali, for example, "*nyAm*a [is] more dangerous than any other ritual objects, and a man who rests his eyes on female genitals will find his way to the grave or suffer blindness" (Brett-Smith 1995, 208–9). Similarly, in an ethnographic study of sexuality among multiple ethnic groups in Africa, Jean-Pierre Ombolo remarks that "genitalia, especially that of a woman, is considered among Africans to be an extremely dangerous reality: it is capable of creating life and supremely capable of destroying it" (1990, 95).[17] By revealing their tabooed body parts or threatening men with menstrual cloths, accompanied by incantations of curses, the women are said to be unleashing the forces that reside therein, causing their male targets a myriad of misfortunes, including impotence, infertility, incurable diseases, and literal or social death. Thus, Malians attributed the 1999 death of dictator Moussa Traoré to a "genital curse" by an elderly woman who cursed him for the shooting of her two grandsons by soldiers loyal to him.[18] Such effect is described as a revocation of life (Turner and Brownhill 2004) and as showing a targeted male his "exit door" (Washington 2005). For the offended women and the community, the collective gesture constitutes the ultimate weapon they wield in calamitous circumstances to punish male targets or to ward off evil, the way warriors deploy their deadliest weapons. Exposing tabooed body parts often makes men listen or comply when other mechanisms of resistance or retribution have failed. Thus, cursing is a form of purification because it rids the land of entities and forces that trample its paramount values.[19]

In the Igbo context, for instance, the angry exhibition of mature female nakedness constitutes the last stage in genital shaming and cursing. Earlier stages include the "verbal suggestion 'Do you want to see where you came from?' [and] the expressive movement (untying the waistcloth in a public and ceremonious gesture)" (Bastian 2005, 46). What exactly the women are pointing toward (the womb or the genitalia) is controversial. Shirley Ardener (1973) has suggested the genitalia, Catherine Ifeka-Moller (1975) has posited the womb, and others such as Laura Grillo (2018) have designated both. Usually, these preliminary steps suffice to deter even the most aggressive male.

Although they are both forms of punishment and resistance, *genital cursing* and its less potent form, *genital shaming*, are distinguishable. Whereas cursing consists of mobilizing otherworldly entities to bring death or misfortune upon a male entity, shaming calls the community to alienate a law-breaking individual or group. Given their powerful implications and consequences for both the woman and her target(s), these gestures are not readily deployed; both are used only in especially deplorable circumstances, in particular wife-beating, insulting women, and arbitrary lawmaking.

To lift the curse or shame and to be reintegrated into the community, offenders must follow a strict procedure: acknowledge their wrongdoings, repent, and plead to be given the opportunity to make amends. These amends include making animal sacrifices to propitiate the dead and spirits whose normative codes have been desecrated, and to purify the shamed and cursed bodies. Among the Beti, for example, the process necessary for the annulment of a curse on a son is as follows: during the public ceremony attended by relatives, the cursed son, who has already brought offerings in preparation for the ceremony, kneels and moves toward his mother in that position. After considering the son's plea for mercy, the mother then goes away to cut a handful of different plants, which she then chews and spits the juice of over him. This ritual reintegrates the son into the community (Ombolo 1990).

As custodians of the social order, women in these societies are tasked with the survival of the community as a whole. They are expected to right its wrongs, which may include punishing offenders, whether of military, political, or royal rank. The women's status challenges the framing of defiant disrobers as mere protesters because their gesture of nakedness is seen as a counterbalancing strategy for the prosperity and survival of the community.

Although insightful and generative, the emphasis on the cosmological can, ironically, be epistemologically counterproductive in that it frames events of aggressive disrobing as mostly empowering for the women. This postcolonial framework is often concerned with and anxious to correct the disempowering images in local and global media of African women as victimized by their traditions. In considerably complicating these commonplace images, traditions become the site of power's actual enactment, not the perceived occasion for oppression. Using ethnographic data produced primarily during colonial and postindependence eras, researchers write as if meanings of the disrobing gesture are often not challenged by the world outside that of the women. In his article "Women's Aggressive Use of Genital Power in Africa" (2006), an analysis of the women of the Niger Delta's threats to go naked in their standoff with the multinational oil company Chevron Texaco in 2002, cultural anthropologist Phillips Stevens Jr. links contemporary events to the mythical and occult aspects attached to the practice, and frames women's defiant disrobing as "genital power." Although a compelling analysis, the article fails to question the uses and abuses that were made of women's images in the international media. Referring to the 2011 Ivorian women's public disrobing, Grillo (2013) also labels their act of disrobing as "genital power" and invokes centuries-long African cosmogonies. In her previously mentioned monograph, Grillo has nuanced her analysis.

As used, such terms predominantly negate a crucial aspect of positions of resistance and power: their temporariness. Although neither Stevens nor Grillo offers a rationale for their choice of terminology, it appears that they attempt to translate into English the practice of defiant disrobing without considering how their usage may reify women's bodies as a space that indefinitely holds something uncritically called "power." To clarify, the word *power* itself is not the problem here; rather, the problem resides in the assumption of the unchanging nature of locations and effects of power. Most importantly, these terms and the analyses from which they emerge seem oblivious to the backlash that modern-day insurgent disrobing encounters both on the continent and in international media.

The uncritical examination often results in internal contradictions. In her analysis of the "mystical actions" of the urbanized associations of priestesses in the Casamance-Senegalese state conflicts, Irene Osemeka (2011) problematically describes the women's protest as peaceful and nonviolent. This framing betrays a certain superficial interpretative work. How does one articulate cursing as a peaceful demonstration when the protesters hope to inflict a certain kind of injury on the body politic? As I show below, Assane Seck adopts a similar problematic strategy in *Sénégal: émergence d'une démocratie moderne, 1945–2005: Un itinéraire politique* (2005).

Such attempts by scholars, which I call the "romanticization framework," seek to uphold ideologically certain images of African women as powerful and their modes of contestation as uniquely celebratory. More recent studies such as Grillo's *An Intimate Rebuke* (2018) provide a more nuanced approach to nakedness as a form of political speech. In her book, Grillo foregrounds the "strategic essentialism" that mature women mobilize in the face of disempowering and imported gender ideologies. This acknowledgment demonstrates that the power effects of FGP can no longer be consigned to a timeless tradition.

Unlike Grillo's study, however, most accounts of uncivil self-exposure give us an incomplete picture of the dynamic that undergirds the performance or threat of defiant disrobing in the postcolonial biopolitical space, where forms of resistance are putatively futile. Given that their societies are becoming increasingly precarious, and in a time when both women and men are experiencing the effects of processes of gender restructuring, precolonial and/or nativist explanatory matrixes offer insufficiently clarifying elucidations. Further, in light of the ensuing intense debates in politics and various media, we need a broader understanding of the constraints and freedoms that produce, enhance, or curtail the power effects of women's insurgent self-exposure. Such is my aim in this book.

If accounts of genital cursing in Africa draw on the local without pretension to universality, reflections on defiant disrobing among dominant Euro-American thinkers invoke provincial contexts for meanings that are clothed in universal garbs. I suggest that although aggressive self-exposure is a universal gesture, it is also one of the most highly context-driven modes of dissent and vulnerability—a classic case of the principle of "universalism without uniformity" that Richard Shweder (2012) has conceptualized. Insurgent nakedness for purposes of purification and cursing is not an African specificity, as countless examples from ancient Greece to contemporary China suggest. A few representative cases include the ancient Greek figure of Baubo (Chausidis 2012; Cohen 1997; Marcovich 1986; Murray 1934); women during the French Revolution (Cameron 1991; Hunt 2002, 2013; Landes 1988, 2003; Vallois 1992); Machiavelli's story of Caterina, the Countess of Sforza (Hairston 2000; Kalogridis 2010; Miles 1989); the Mexican writer Salvador Novo's *La guerra de las gordas* (1994);[20] Chinese women's ritual self-exposure during the defense of the city in western Shantung during the Wang Lun uprising of 1774 (Henry 1999); and the Irish Sheela-na-gig ritual of genital flashing (Dexter and Goode 2002; Freitag 2004; Rhoades 2010).

Despite its transhistorical and transcultural nature, however, intentional self-exposure can defeat the yearning for endowing what is essentially a universal act with universal meaning. The intended meanings of naked self-exposure often fail to transcend their local connotations. In that sense, I call the gesture a "signifying shorthand" because its meanings not only invite but also demand further exploration.

The Concept

The new theoretical framework that emerges from this study is "naked agency," which designates the incessant negotiation of power relations between the women and their targets and other stakeholders. This dynamic can best be understood in terms of openness, or figurative nudity, of naked agency. This concept alludes to the unsolicited yet generative encounter between African local cosmologies about exposed tabooed skin; Africanized political institutions; and dominant accounts of nakedness as a state of vulnerability, truth, and innocence. More broadly, my work also disturbs the commonplace images—promoted by the media—of rape, mutilation, and pathology associated with women's bodies in Africa. It also fundamentally reframes questions of women's sexuality and power by offering an alternative to opposition of sovereign subjects and their victims that informs much research in African studies.

Given the ubiquity of discussions of female genital cutting (FGC), an act that emerges in the transnational discourse on African agency and gender politics, readers who are interested in African women's sexual repertoires might be tempted to place questions of FGC into the discourse. In fact, the in-between space that women who display naked agency inhabit is indeed a space similar to that occupied by women in contexts of FGC. As Chantal Zabus (2007), Fuambai Ahmadu (2017), Wairimū Ngaruiya Njambi (2004), Lori Leonard (2009), and others have argued, there is a kind of agency in local African women fighting against, or endorsing, FGC, a long-established tradition that some deem harmful. This manifestation of agency both curtails and amplifies male power. Female secret societies, the agents of genital cursing, are also customarily the guardians of the "excision" ritual among young girls. The power effect of the clitoris explains why it is feared, ritually worshipped, and cut as the paramount punishment for adulterous women, as Jean-Pierre Ombolo (1990) notes in his comparative anthropological accounts of sexuality in Africa. These resistant practices, both progressive and regressive, regarding FGC emerge in contexts caught between two putatively opposed orders, authenticity and foreign influences, as Saida Hodžić argues in *The Twilight of Cutting* (2017).

In one sense, defiant disrobing is not fundamentally antipatriarchal; rather, it works in tandem with patriarchy. The gesture derives its power effects from the patriarchal parameters of anatomical determinism—worth emphasizing in light of interpretations of these gestures as implying safety, freedom, and success for the women. These considerations further nuance the account of agency that this book promulgates.

Naked agency is more capacious than any single indigenous African term because each ethnic group examined has its own terminology (table I.1). All these terms describe female-only actions. The term *adjanou* in Baoulé does not translate as *naked*, or *nakedness*; instead, it refers primarily to the sacred and secret exorcism ritual during which naked elderly women mobilize the entities within their bodies either to curse an offender or to ward off evils. Similarly, the name *anlu* in Kom to designate genital-cursing rituals actually means "to drive away." Within that name resides a reference to a famous Kom legend in which women disguised themselves as men to drive away the neighboring Mejang. *Adjanou* and *anlu* are not just descriptive of nakedness; in Ivorian and Cameroonian representations, they evoke fear and awe because they carry historical and customary connotations as well as unexplored and complex valences. Imposing a single term would flatten geographical and historical specifications about the meaning of public disrobing. The term *naked agency* accomplishes the task of moving beyond the ethnic and the local without carrying specific ethnic connotations but also allowing for them.

Unlike terms such as *genital cursing*, FGP, and *oto*—which suggest fixity, localization, the ethnic, and freedom from the effects of historical and social variations—*naked agency* names a complex and unstable gesture. Given the slippages, confusions, and often contradictory motivations and responses, the women's gestures are a set of "open strategies"—naked strategies, with positions that are constantly subjected and emerging.[21] Although this study's starting point is the African context, I ultimately move beyond that continent to name and highlight the dominant features of most instances of resistant self-

TABLE I.I. Ethnic terms in sub-Saharan Africa

Term	Ethnicity	Country
adjanou	Baoulé	Côte d'Ivoire
anlu	Kom	Cameroon
bomampi	Attié	Côte d'Ivoire
ebgiki	Abidji	Côte d'Ivoire
gbiteté	Ewe	Togo
guturama	Kikuyu	Kenya
kilipat	Pokot	Kenya
koo	Bassa	Cameroon
mevungu	Béti	Cameroon
momomé	Agni	Côte d'Ivoire
moribayassa	Malinké	Côte d'Ivoire, Guinea, Mali
ndong	Balong	Cameroon
olkishoroto	Maasai	Tanzania
oto	Igbo	Nigeria
setshwetla	Zulu	South Africa
titi ikoli	Bakweri	Cameroon

exposure: the fluctuation between positions of victimhood and sovereignty, or more accurately, victimhood in sovereignty, or sovereignty in victimhood.[22]

In modern usage, although *naked* and *nude* are often used interchangeably to designate the unclothed human body, they are encumbered with cultural variations and historical determinations. Definitionally, *naked* derives from Germanic *nackot* and means "having no clothing on the body, stripped to the skin; unclothed."[23] In Middle English, *naked* moved from being merely descriptive to becoming evaluative—destitution, and shame in the naked person. It was then frequently used in "the context of a person's birth, perhaps to connote a newborn child's vulnerability or innocence" (OED). This connotation has persisted into the present. *Nude*, from the Latin *nudus*, was primarily used in the legal context to mean "not attended by any formalities or pledges," or "lacking consideration."[24] *Nude* also meant "open, simple, plain, naked, bare, unclothed" (OED). Whereas *naked* is used more generally to designate vulnerability and absence of normal clothing, *nude* is utilized to refer to nakedness considered positive or aesthetic, especially in photographs and other works of art. These associations reflect the implicit relation to rank and hierarchy that characterizes Anglo-Saxon and Latin words (Barcan 2004).[25] And that negative conception of clotheslessness is automatically opposed to that of agency, self-determination, and resistance (Agamben 2011; Nancy 1993).[26]

Naked in *naked agency* makes sense when we perceive the amplificatory impact of the imbricated categories of gender and race on exposed genitalia. Surprisingly, though, the women can be both naked and nude. They can be described as *nude* because they are clothed in cultural significations: in both the grammars of reverence from their perspectives and those of uncivility from the viewpoint of their targets and opponent-critics. However, considering the historical discursive injury that has cast images of African female bodies as vulgar, with its residual effects that continue to shape current valuations of these images, the Nigerian women's bodies during their 2002 threat may be discursively labeled *naked*. Although the term *nude* suggests multilayeredness and may break away from the long, historical wound inflicted on black bodies, I found its connotations too artistic and tame for the kind of political and social work that the protesters are doing. Either way, these unclothed bodies escape the rhetoric of plainness, understood as meaninglessness. Here, *naked* is used both figuratively and literally to emphasize the confrontational exhibition of bodies in public as well as the women's lack of options.

Agency in *naked agency* designates women's ability to act or react, intentionally or otherwise, in punishing offending males or signaling vulnerability, or both. My use of "agency" disavows the dominant liberal model that is equivalent to autonomy, intentionality, free will, sovereignty, and transcendence. Not surprisingly, these liberal accounts of agency, provincial and yet clothed in the garb of the universal, are themselves being resisted both by practices of agency and by theorizations of agency. *Naked agency* is one such theorization that calls attention to cultural and geographical corners, as well as to certain peoples who have been left outside the original history or seminal conceptualizations of agency. I engage but exceed each of the four dominant conceptualizations of agency in poststructural and postcolonial circles. Whereas Gayatri Chakravorty Spivak (2009) defines agency as institutionally validated action, Gabriela Basterra (2004) formulates it as the power one feels. These two versions draw on Michel Foucault's classic reflections on power (1978, 1979, 1980, 1982, 1997) and Judith Butler's account of the paradoxical nature of subjection in *The Psychic Life of Power* (1997). Through disobedient nakedness, the actors may be simultaneously negotiating between institutional validation and survival instinct. It is unclear whether Nigerian women's targets drew on customary belief systems to interpret the women's threat to disrobe. However, given that the women's ability to react, which succeeded in demonstrating their agency, backfired in some ways, it becomes clear that there is no agency outside of restrictions. This interpretation of *naked agency* exposes important aspects of what it means to be an acting agent in challenging material conditions.

My focus on naked agency should not be misconstrued as an argument for defiant disrobing as the only avenue of collective female resistance. It should not be conflated with an uncritical excavation mission. The spirit of this book, which is to transcend dichotomies and "to learn from below," resists such monolithic representational politics. Because of today's higher female literacy rate, women in Africa marshal a greater variety of conventional tactics of participatory democracy. Even those who expose themselves, either in anger or in ritual, often mobilize conventional strategies to prosecute grievances or to bring about change.

The Scenes

Throughout the book, I use the term *scene* rather than *chapter* to foreground the performative nature of the women's act, thus joining the performers and their audiences into a mutually constitutive dynamic. The scenes are organized into three sections: (I) Restriction, (II) Co-optation, and (III) Repression, each highlighting a different aspect of naked agency. Table 1.2 shows, alphabetically by country, the notable female naked protests over the last nine decades that I will be drawing on in my discussions.

Section I, "Restriction," unpacks sociopolitical circumstances that require the women's gesture of disrobing at the same time that they constrain the extent of the women's agency.

Scene 1, "Exceptional Conditions and Darker Shades of Biopolitics," analyzes the 2011 women's march and cursing ritual in Côte d'Ivoire to highlight the once-exceptional but now normalized conditions that call for disrobing in contention. These conditions, without which the women's ability to act would have been invisible, have been explained as biopolitical and even necropolitical. Biopolitics is the imbrication of life processes into political calculations in the name of enhancing peoples' lives. The inherent contradiction of a biopolitical form of power is that it creates a scheme in which some people enjoy political and legal prerogatives and protection while others are depoliticized through the normalization of the state of exception. Similar contexts evince the possibility of emancipatory political subjectivities, which carries broader implications for biopolitical theorizing by decentering its dark side.

Scene 2, "Dobsonville and the Question of Autonomy," closely reads *Uku Hamba 'Ze (To Walk Naked)* (1995), a short documentary on the 1990 female naked protest in South Africa, to uncover the texture of women's self-determination. With access only to the women's regrets and triumphs, I argue that the actors' agency is precarious because of their extreme emotions—fear,

TABLE I.2 Notable female naked protests in sub-Saharan Africa, 1922–2017

Country	Year	Protests and Grievance(s)
1 Benin	2013	Women of the market of Gbogbanou: threat against market clearance
2 Cameroon	2010	Farmer and grazer land conflicts
	1993	Takumbeng: electoral injustice, dictatorship
	1958	Kom women: cross-contour farming issues
3 Central African Republic	2015	Young women: rape
	2014	Women: sectarian violence
	1979	Mothers: Bokassa's dictatorship, arbitrary detention
4 Côte d'Ivoire	2011	Paris: international entanglements
	2011	Abobo, March: electoral injustice
	2011	Yamoussoukro, February: electoral injustice
	2010	Didievi: Gbagbo's dictatorial practices
	2003	Exorcism, Dominique de Villepin/neocolonialism
	1949	Political prisoners
5 Democratic Republic of the Congo	2014	Factional politics, support of Alex Kande Mupompa
	2012	Refugees in Uganda, better living conditions
	2006	Killings of elderly women
6 Gabon	2015	Market women: racketeering
	2009	Election-related contestation
7 Gambia	2001	Sacrifice of a dog for electoral purposes
8 Ghana	2007	500 women: threat to march naked to the seat of government against dismissal of district chief executive
9 Kenya	2001	300 women: land dispute, wildlife preservation programs
	1992	Political prisoners, Wangari Maathai
	1922	Harry Thuku arrest protest
10 Liberia	2014	Widows: payment of benefits
	2008	Refugees in Ghana, relocation packages
	2003	War negotiations, Leymah Gbowee
11 Mali	2012	Rumors of women walking naked against Amadou Toumani Touré
	2003	Dictatorship, military violence

(*continued*)

TABLE I.2 (*continued*)

	Country	Year	Protests and Grievance(s)
12	Nigeria	2016	Kaduna, December: against Gov. El-Rufai
		2016	Ebonyi, October: detention of community president-general
		2015	Ibadan, November: white elephant project
		2014	Kaduna, September: governor's visit, killing fields
		2013	Threat against child marriage bill
		2012	Ogun: invasion of the community by hoodlums
		2009	Ekiti: electoral injustice
		2003	Electoral injustice
		2002	Niger Delta: multinational oil companies' exploitative practices
		1947	Yorubaland: female taxation
		1929	Igboland: the Women's War, female taxation
13	Republic of the Congo	2002	Electoral injustice
14	Senegal	1980	Usana: students' strike
15	Somalia	1996	Dictatorship
16	South Africa	2016	university students, October: "Fees Must Fall"
		2016	university students, April: sexual harassments and rape
		2016	Pretoria: local elections
		2012	Limpopo: water shortages
		2006	Prison inmates: relocation
		2001	Kempton Park in the Gauteng province: women's protest against police for denying them access to Bredell farm for their belongings
		1990	Dobsonville: slum clearance
		1959	Beer hall boycott
17	Sudan	2009	Dictatorship, men's inactivity against dictatorship
18	Swaziland	2000	Land dispute
19	Tanzania	2003	Land dispute
		2000s	Criminalization of female genital cutting
		1977	A government decree that Maasai must wear "modern" dress to use public transportation

TABLE I.2 (*continued*)

	Country	Year	Protests and Grievance(s)
20	Togo	2013	Manipulation of constitution, dictatorship
		1933	Increase in taxes and the levying of new fees on market women; arrest of two local political leaders
21	Uganda	2015	Acholi women: land dispute
		2012	Police brutality
22	Zambia	2017	Mothers and grandmothers: topless to express their hurt at government's detention of their leader, Hakainde Hichilema
23	Zimbabwe	2016	Threat of disrobing against police brutality

desperation, anger—and social and historical determinisms. Despite standing up to the country's apartheid regime, these women fluctuate between history-making and patent victims. That in-between position, what Carole Boyce Davies has called in a different context "the migrating subject" (1994), also resonates with the protest as standing at the junction of apartheid and postapartheid eras, between the 1959 beer hall boycott and the recent cross-ethnic shaming nudity deployed by South African students. Exploring this lineage demonstrates that the meanings of resistant nakedness (available in European languages and circulating internationally) differ markedly from the account of "genital cursing" as annihilating offenders by calling on ancestors in West Africa. The comparison challenges the concept of Africa as a unified entity.

Section II, "Co-optation," foregrounds complex (disabling and enabling) responses to the women's gesture. The lauding reactions paradoxically magnify and undermine women's agency, thereby highlighting the dialectical nature of insurgent disrobing.

Scene 3, "Africanizing Nakedness as (Self-)Instrumentalization," draws on historical accounts and newspaper articles from the late 1950s to the 2010s on Côte d'Ivoire, Cameroon, and Senegal. I show the ways in which civic leaders Africanize, read "overplay," the power effects of resistant nakedness for nationalist or factional political interests. However, by exploiting the social capital of tradition and indigenous religious practices to become agents while being exploited, the women problematize current and simplistic instrumentalization arguments that emphasize women as victims of powerful forces.

Scene 4, "In the Name of National Interest," engages the 2008 arrest of Liberian women war refugees in Ghana and the shame dance in Ngũgĩ wa Thiong'o's fantasy novel *Wizard of the Crow* (2006). This scene showcases a different kind of exploitation that works in two ways. The first aspect allows postcolonial bureaucrats to use mature female insurgent self-exposure to silence dissenting voices and to carve out a space of agency for themselves. The second aspect consists of civic leaders' co-opting "traditional" women's dances for "national" purposes, and the women's own instrumentalization of genital flashing—not to advance economic interests as in scene 3 but rather to shame elected officials. The goal of this scene is to uncover intentional self-exposure as both enabling and repressive for civic leaders and the women themselves.

Scene 5, "Film as Instrumental and Interpretive Lens," continues the thread of co-optation with a discussion of the postmodern film *Les Saignantes* (2005), demonstrating how the filmic text exploits women's purifying Mevoungou ritual to heal the postcolonial modern state. The Mevoungou ritual—banned by German missionaries in the nineteenth century—is revered among the Beti of Cameroon because its power centers on the clitoris, which women worship in moments of communal crises to ward off evil forces. I closely read the film to underscore not only the apocalyptic vision of a necropolitical state but also the necessity of enlisting a female-centered ritual to heal the ills of the found(l)ing fathers. Through women's bodies and the film's tropological counterdiscourse to postcolonial necropolitics, hope in the deathscapes becomes viable against the further implementation of dystopia.

Section III, "Repression," documents the physical and rhetorical backlash that defiant disrobing encounters within postcolonial biopolitical circumstances. This backlash problematizes arguments of genital cursing or female genital power as uncritically empowering.

Scene 6, "Secularizing Genital Cursing and Rhetorical Backlash," explores news reports in several countries to highlight what I call the "secularizing" tendency—efforts by public officials and intellectuals to strip the gesture of its assumed religious connotations, and of its harming and murderous effects. Consequently, the gesture, which becomes a violation of decorum within political secularism, also becomes available for state repression. The secularizing/deritualizing trend—the most widespread of all reactions—indexes the unfinished business of the postcolonial nation-state regarding its indigenous religious practices. Yet to consider deritualization as only constraining is to disregard how it benefits women protesters, including South African students of the Fees Must Fall movement, who are not banking on the sanctity of their bodies or on their status as moral guardians to inflict shame.

Scene 7, "Epistemic Ignorance and Menstrual Rags in Paris," explores a manifestation of the secularizing trend and draws on recent social-media material and newspaper reports in which Ivorian women cursed American and French political authorities. My goal is to demonstrate how these ritual cursings, called Opération Kodjo Rouge, and silence, a form of epistemic ignorance from the French political class, contribute to a deeper understanding of how globalization both reads and represses its own colonial histories. Globalization has enabled the visibility of specific social movements. Yet, given the standardization of cultural, political, and social norms, globalization also represses modes of political dissent not consonant with the Enlightenment-inflected and bourgeois-informed channels of participatory democracy. Although visible, the modified women's cursing rituals, the most violent gesture they could unleash, are considered nonviolent and are therefore ignored or dismissed in a paradoxical dynamic of globalization.

Scene 8, "(Mis)Reading Murderous Reactions," closely examines T. Obinkaram Echewa's historical novel *I Saw the Sky Catch Fire* ([1992] 1993), a fictionalization of the 1929 Women's War, in which thousands of Igbo women exposed their naked buttocks. Here, I account for the murderous responses that self-exposure has met in the colonial period. In rewriting the colonial gendered trauma through specific formal strategies, the novel provides the most nuanced account of the political stakes of naked agency. The restaging of the war displays a wide range of instances of aggressive disrobing in order to highlight both the protesters' and the targets' emotions and counterattacks. In that sense, the novel displaces the commonplace images of victimhood attached to women's bodies without falling into the easy trap of triumphalism.

The epilogue, "Defiant Disrobing Going Viral," revisits the question of proliferation first addressed in the introduction. Despite its proliferation, and as a signifying shorthand, naked protest is subject to translation and mediation, and its performance demands further exploration. As the most universal and yet the most highly context-driven mode of dissent, the deployment and reception are often predicated upon a host of ways of thinking about the body, sexuality, privacy, moral injury, clothing, and other factors that transcend social sharedness. I conclude by asking whether it is reasonable to wonder if, as more and more bodies get naked in both real and virtual spaces, that which is the ultimate weapon for some women might become merely one more ludic spectacle of globalization.

Ultimately, I hope these scenes cumulatively will enhance our understanding of forms of contestatory agencies in a world of increasing precarity, neoliberal biopolitical practices, and globalization. The negative tonality in current biopolitical thinking as a response to feelings of extreme perishability has led some to consider pointless any form of resistance, with the consequential degraded view of political subjectivity. In bringing to our attention these protesting women and their exposed bodies, *Naked Agency* argues that desperation and vulnerability in exceptional circumstances can constitute a generative category for deliberating on political subjectivity.

Section I

Restriction

Exceptional Conditions
and Darker Shades of Biopolitics

I strongly condemn the abhorrent violence against unarmed civilians in Côte d'Ivoire. I am particularly appalled by the indiscriminate killing of unarmed civilians during peaceful rallies, many of them women. . . . On March 8, the 100th anniversary of International Women's Day, we saw pictures of *women peacefully* rallying with signs that said, "Don't Shoot Us," a strong testament to the bravery of women exercising their right of *peaceful assembly*. (emphasis mine) —**Barack Obama**, White House press release, March 9, 2011

Perhaps, if Foucault could have seen the way African "demography" is "regulated" by the AIDS endemic (and a number of other epidemics, all monitored by a "World Health Organization"), he might have ventured to speak of *"negative bio-politics."* (emphasis mine) —**Étienne Balibar**, *Politics and the Other Scene*

Recalling the circumstances that require or even demand women's use of their nakedness in protest highlights the necessity of conceptualizing both their desperation and their courage. The intellectual purchase of this conceptualization makes more sense when we understand our era as one in which the individual and the community are alienated from normative structures (values, institutions, legislations, and practices) and exposed to illiberal arrangements under the pretext of exception and crisis—conditions that have led many to decry the disappearance of human political agency. However, paying attention to non-Eurocentric contexts suggests the ways in which women in Africa are carving temporary spaces of contestation using resistance techniques that dominant conceptualizations of sociopolitical agency and participation have not fully considered.

On March 3, 2011, in Abidjan, Côte d'Ivoire, against repeated presidential orders banning protest demonstrations, thousands of desperate but empowered women took to the streets. Some even exposed their nakedness to curse the incumbent president, Laurent Gbagbo, in order to oust him from office. Gbagbo's refusal to cede power after the long-awaited November 2010 presidential elections, after which Dr. Alassane Ouattara was declared the winner by both the United Nations and the African Union, brought the country to the brink of civil war. Businesses closed, the economy died, and the country became isolated from the rest of the world. For more than three months, rebel soldiers loyal to Ouattara exchanged gunfire with national defense and security forces.

The protesting women were desperate because the long series of protests and subsequent murderous repressions during the protracted postelectoral crises left impoverished populations to bear daily scenes of terror and deprivation, with no access to electricity and running water. Yet they were empowered because as mothers with the prerogatives thereof, they considered themselves responsible for the survival of their society. Their self-exposure was meant to purify the land of those who trampled its normative values. Thus "armed" with the locally understood and authorizing symbols attached to their status as mothers, and carrying representatives of nature such as leafy branches, pestles, and other cooking utensils, the women invaded the dangerous public sphere.

The all-women's march soon took a terrible turn, however. In an amateur YouTube video broadcast by CNN and CBS, and thus attracting the world's attention, the women's singing was interrupted by a sudden burst of gunfire that spewed from a mounted machine gun. In the ensuing mayhem, the fragmented body parts of some women were spread and intermingled, thus erasing singularity. Other women fell down and were trampled. Six women were killed on the spot, a seventh died in the hospital, and nearly one hundred were wounded as the streets of Abidjan ran with blood.

News of the carnage caught the attention of the "leaders of the free world": U.S. Secretary of State Hillary Rodham Clinton and President Barack Obama immediately condemned the shootings of "unarmed" women. A few weeks later, on April 10, the United Nations Operation in Côte d'Ivoire (UNOCI), supported by French Licorne forces, launched a military operation in Abidjan in retaliation for the Gbagbo forces' use of heavy weapons against civilians and UN peacekeepers. This operation ultimately led to Gbagbo's surrender on April 11. In January 2019, the president was acquitted (pending appeal) of crimes against humanity by the International Criminal Court (ICC) in The Hague in the Netherlands.

In most postcolonial African countries, the nominal state is often the major instigator of scenes of violence that take place during the normalized state of exception, what I call "sacralizing" conflicts. In these conflicts, not only is the biological life of the average citizen in constant danger of annihilation but also her political life is actively undermined by the manipulations of constitutions or the postponement of presidential elections. The brutal and murderous repression makes any form of contestation a matter of life and death.

That security and defense forces loyal to Gbagbo used war-grade weaponry, including fragmentation grenades, against their citizens shocked both Ivorians and outside observers. Morgues were full, and body bags arrived "dripping with blood" (Callimachi 2011). The heavy artillery used in the deaths of the women included "a green pickup with a mounted machine gun, a police cargo truck, a green military camouflage tank, and a blue gendarme tank" (Human Rights Watch 2011). The shocking indication to Ivorians that the military was ready to use war-grade weaponry was confirmed by aid workers and organizations. "A 14-year-old's corpse had hundreds of shrapnel wounds across the chest, and the doctor who attempted to save him last week said the wounds were the result of a fragmentation grenade, similar to those used in Iraq and Afghanistan," reported Rukmini Callimachi (2011). Most news reports in *The Guardian*, the *New York Times*, the *Washington Post*, and other sources emphasized the shocking realization by the local population that Côte d'Ivoire had in fact entered the state of exception, more violent than the one experienced during the 1950s struggles for national liberation. Could it be that the postcolonial era is turning out to be more lethal and brutal in some ways than the colonial?

During the colonial period, demonstrations of this kind, including incidences of genital cursing (the women's term), by women in Côte d'Ivoire were dispersed with milder forms of repressive tactics such as detonating teargas grenades and firing shots into the air, as already argued. The specific nature of postcolonial violence resides in the fact that, unlike forms of colonial governance built on exploiting both human and natural resources, modernized regimes that are putatively invested in securing the citizens' lives are instead maiming and killing women in the name of nationalism.

The March 3 march/cursing-ritual-turned-carnage is a story of many facets: desperation, power, deadly assumptions, inconsistent explanations, crucial misreadings, and political co-optation. Its retelling is not to satisfy a morbid attraction to death and bloodshed but rather to suggest the kinds of exceptional uses of state violence in the postcolonial era that beget assaultive nakedness or genital cursing. This performance of desperation and the ability to contest, often in the form of instinctive resistance via the revered and stigmatized maternal body, constitutes a blueprint for analyzing the nature of resistance. This blueprint is visible in the generalized African context and, by extension, in the world that paradoxically generates, and is exposed to, multidirectional forces. Expanding Judith Butler's (2006) attempt to conceptualize political deliberation amid aggression and injury, I revisit the enduring question of the agency of historically marginalized subjects and their largely discounted forms of political participation. Reading the performance of this unconventional form of oppositional politics may provide insight into current debates around gender, customary beliefs, and active resistance (or its impossibility) in the negative biopolitical frame. Analysis of defiant disrobing challenges theories of biopolitics that argue for the futility of any forms of resistance in the face of impending death and of omnipresent vulnerability.

Darker Shades

Related to Foucault's genealogical work on prisons, sexuality, and governmentality, biopolitics emerges from his often-incomplete reflections on the rise in nineteenth-century Western Europe of various forms of knowledge and practice resulting from the implementation of a politics of population control.[1] The most notable of these phenomena is the imbrication of life processes into political arrangements. One of the major consequences of the shift from sovereign power (to take life or let live) to biopolitics (to make life and let die) is the creation of a sociopolitical scheme in which some lives enjoy political and legal prerogatives and protection while others are depoliticized and effaced. In sum, the concept exposes the paradoxes of modern democracies, in which the technologies of power both enable and disallow life.

Biopolitics is a clarifying concept because of the kinds of insights it offers about current and not-so-distant social, economic, and political climates. However, explaining the ubiquity of death and vulnerability in our era when theorizing about biopolitics has focused on Western democracies, at the expense of other contexts, marked chiefly by the effects of postcoloniality, race, gender, and sexuality (Giroux 2006; Mbembe 2002; Weheliye 2014). Specifically, countries in Africa withstand the worst of unexpected violence and precarity because of geopolitical configurations such as neoliberal policies and the painful afterlives of colonialism. Neoliberal policies and their obsessive investment in the security and improvement of the life of the population paradoxically often necessitate the state of exception, suspension of normative structures, or the creation of deathscapes—contexts in which populations are reduced to the status of living dead (Mbembe 2003, 40).

In Africa, the entities involved in these creations include war machines, the state, the political class, and multinational corporations with avaricious global entanglements. Several observers have seriously implicated the opposition in the shooting of the women. Reportedly, opposition leaders knew very well that the women would be murdered. Based on that information, they masterminded the rituals. The opposition's plan was to prompt the intervention of the French army and the UN coalition force in the Ivorian political landscape (Varenne 2012). Thus, I look at the demonstration and massacre to argue that the opposition used women as pawns against the regime of the incumbent, Gbagbo, given aid donors' sensitivity to images of oppressed "Third World" women. The invitation of Ouattara's coalition with the Parti Démocratique de la Côte d'Ivoire (PDCI, Democratic Party of Côte d'Ivoire) to women to demonstrate, putting forward the rhetoric of sacred motherhood and the mother figure as the guardian of the nation, was a misguided and desperate attempt, without any consideration for the women's lives (Salyff G. 2011; Varenne 2012).

Debates continue about the identity of those who fired, whether they were from Gbagbo's camp or if they were mercenaries hired by the opposition to smear Gbagbo's image. Independently of those questions, the fateful day demystified the rhetoric of the sacred motherhood that has been the foundation of the Ivorian nation-state. As the videos of dismembered women's bodies were going viral, the image of Gbagbo as a ruthless dictator was being stamped on the consciousness of the international community, thus sealing his fate. I read these women protesters as figures who are denied political recognition and yet are constitutive of political intrigues. Paradoxically, the protesters' deaths became instrumental in writing them into history and politico-juridical battles; their massacre was included in the charges against Gbagbo at his trial in The Hague.

In addition to public officials, proliferating "war machines" (Mbembe 2002), to use Mbembe's term, senselessly (it seems) strike defenseless groups and individuals. These machines include rebel groups, Boko Haram in Nigeria and neighboring countries, and the "microbes" in Côte d'Ivoire (Adélé 2016; Hamid 2016). They fashion the fear of extinction in locales with no declared civil war, blurring the distinction between war and peace. In 2014 Kaduna, a state in northern Nigeria, experienced numerous and repeated massacres, leading some to call Kaduna the "killing fields." Mukhtar Ramalan Yero, governor of Kaduna State, declared, "Life has become so cheap in the North. Innocent people are killed on a daily basis without justification" (quoted in Ehiabhi 2014). The outraged governor then solicited the assistance of the community in inappropriate ways by asking them to confront the heavily armed mass killers. This declared inability of the state to ensure the security of citizens angered local women, who stripped naked to decry the governor's negligence and poor leadership.

When local war machines do not cause numerous deaths of civilians, imported toxic pollution does. The unloading by the ship *Probo Koala* of several hundred tons of dangerous chemicals in Côte d'Ivoire in 2006 has significantly harmed communities in the area (Bohand et al. 2007). This case of toxic pollution, which led more than 100,000 Ivorians to seek medical attention (Voice of America 2006), is an overt manifestation of global connectedness and greed. The *Probo Koala* is registered in Panama, chartered in 2006 by the Dutch-based oil trader Trafigura, which in return entrusted the waste to a state-registered Ivorian company, Tommy. This dumping fits into a larger account of the continent as one populated by the living dead (Mbembe 2003). A 1991 alleged internal memorandum by then vice president and chief economist of the World Bank, Lawrence Summers, suggested, in a modern Swiftian Modest Proposal, the dumping of toxic wastes in the "UNDER-populated" and "UNDER-polluted" countries in Africa (*The Economist*, February 8, 1992).[2] The memorandum stated that outrage over a possible pandemic of prostate cancer that the waste might cause would be unfounded because poor people in African countries do not live long enough to develop prostate cancer.[3]

If imported toxic waste does not sufficiently impair the lives of poor populations, the mismanagement of pandemics should do so. Management of HIV/AIDS at the peak of the pandemic involved a host of actors, ranging from denialists and saviors to offenders and sufferers. The major denialist was the former South African president Thabo Mbeki, whose AIDS denial "caused 300,000 deaths" (Boseley 2008; Dugger 2008; Mbeki 2016; Roeder 2009).[4] Well-meaning peddlers of volunteerism in Africa engage in activities akin to dark tourism—travel to places historically associated with death and tragedy. Avaricious unethical pharmaceutical firms and their complicit African public officials exploit human and natural resources through mismanagement of public resources and petty, bureaucratic, and political corruption. For instance, following Trafigura's scandal, in 2007 the company agreed to an out-of-court settlement of $200 million with the Ivorian government. A court ruling also required Trafigura to pay $45 million to thirty thousand alleged victims. However, by January 2010, few individual claimants had received cash compensation. Eventually, investigations demonstrated that public officials had embezzled parts of the funds (BBC News 2012; A. Konan 2012).

Against these figures are overwhelmed local medical workers, and AIDS sufferers and their social-activist advocates, who fight for human dignity and citizenship. In 2002 the circumstances surrounding the AIDS endemic in some

African countries led philosopher Étienne Balibar to conjecture: "Perhaps, if Foucault could have seen the way African 'demography' is 'regulated' by the AIDS endemic (and a number of other epidemics, all monitored by a 'World Health Organization'), he might have ventured to speak of 'negative bio-politics'" (2002, 38). In this conjecture, combined with disappointment and doomed hope, Balibar posits that Michel Foucault could have expanded his theorizing of biopolitics with attention to something called "Africa."[5] I insist on the fictionality of "Africa" because varied populations on the continent differentially experience random violence.

In this landscape of greed, care, and resilience, the HIV sufferer is akin to naked life because "in many places, access to medicine—rather than, say, jobs, clean air, or freedom from war—has come to epitomize citizenship, equity, and justice" (Comaroff 2007). However, Jean Comaroff describes a complex picture, one that includes not only gloom and doom but also the human capacity for contestation and affirmation. She connects such human capacity, which is "beyond bare life," to the emergent forms of political subjectivities that belie the images of abjection that have been conceived to define HIV/AIDS in Africa. In many and powerful ways, these AIDS sufferers and their allies have moved beyond the "self-insulating, self-referential circuits of communication and concern" of academics (Comaroff 2007, 215).

Sacralizing Conflicts

One such creative affirmation is the 2011 Ivorian women's cursing march held in a moment of rampant fear and perishability. The moment approximates the state of exception, wherein the distinction between war and peace remains indistinguishable. To justify his refusal to cede power, Gbagbo and his regime invoked the dignity of the Ivorian civilian body. However, the process of conferring dignity implies the possibility of doing away with civilians' lives in brutal postelection crisis by using terror-producing tactics and fabricating the state of exception through the suspension of most legal, constitutional, and political privileges, including the right to protest. In Côte d'Ivoire, Gbagbo's ultranationalism in his us-versus-them rhetoric in the 2011 postelectoral era literalizes the notion of sacralizing conflicts.

Drawing on Giorgio Agamben's (1998) exploration of the logic of sacralization at the heart of Western political tradition, I use "to sacralize" here to designate a political maneuver whereby sociopolitical leaders, in the name of the sacredness of the population's lives, incite their followers to resist when their lives are unmistakably endangered. Such sacralization serves to generate and secure sovereign power. For instance, during the November 2010 electoral campaigns, one of Gbagbo's posters reads, "Laurent Gbagbo, 100 percent for the Ivory Coast." According to Reuters staff reporter Tim Cocks (2010), "Gbagbo owes his popularity in large part to his skill at projecting himself as an arch nationalist defending Ivory Coast from sinister 'foreigners,' whether they be Ouattara, [the country's] former colonial master, France, or economic migrants from Ivory Coast's poorer, drier neighbours like Burkina Faso or Mali."

Foucault pithily summarizes these tactics in the Western context: "Welcome to the nation, go get slaughtered and we promise you a long and pleasant life" (2000, 405). Gbagbo and his camp's warmongering rhetoric seem to say, "Go get slaughtered and we promise you a free and decolonized nation."

In such protracted conflicts, the average civilian cannot be sacrificed, but she can be killed with impunity by anyone. Naturally and unbeknownst to her, she comes to embody the figure of the *Homo sacer*. The crime of the Ivorian *Homo sacer* is to be born in poverty, and to be positioned on the wrong side of political, social, economic, ethnic, or religious divides. The state of exception and the prevailing circumstances create a sense of widespread vulnerability and desperation. The 1998 founding in Italy of the ICC promises to eradicate the impunity that most postcolonial bureaucrats have enjoyed by seeking to judge war crimes around the world.[6]

Perhaps in other contexts, the sense of perishability could have ended the women's attempt at sociopolitical agency in 2011 in Côte d'Ivoire. However, banking on their cultural prerogatives, the women carved a temporary space of agency in which to gain the attention of the international community. Yet the women made two fatal assumptions, because the postcolonial city is shaped by ethnic diversity, and also because their bodies are coded differently in the postcolonial state on the brink of a civil war. For instance, the organizers and participants in the march/cursing ritual wrongly assumed that even if Laurent Gbagbo's security and military forces shot at men, they would be too ashamed to open fire on and kill women.

At least two major contradictory reasons for the shootings emerge. For one of the leaders of the women's cursing march, the quinquagenarian Mrs. Touré, the carnage happened because "in Africa, and Ivory Coast, [naked or black-wearing mature female demonstrators] is like a curse. That's why the soldiers were afraid and shot at them" (quoted in D. Smith 2011). Surprisingly for the protesting women and for all the observers, genital cursing became murderous, not for the targeted Gbagbo or the security forces who laid eyes on women's exposed genitalia but for the cursers themselves. Paradoxically, the massacre also legitimizes the belief in the efficacy of genital cursing to create fear in men: if the security forces did not share the same understanding of exposed mature female bodies as deadly, perhaps no shootings would have taken place or the soldiers would not have had to show their capacity for retaliatory action.

The soldiers' extreme reactions could have been due to something entirely different from feeling cursed by being exposed to women's genitalia in a ritual setting. Contradicting Touré's explanation for the shooting is one offered by a military source that "security forces believe rebels sometimes hide among civilians. It was a blunder that we regret. . . . It is unfortunate" (quoted in Bax and Smith 2011). The military explanation is a confusing labyrinth of "belief," "sometimes," and "hide," which are markers of firsthand experiences. This second explanation contradicts the first: the soldiers' fear comes not from the actual sight of exposed stigmatized mature female nakedness, which may or may not be a curse, but rather from another force, the belief in the putative presence, in hiding, of the "rebels." Either way, fear is at the heart of the deadly shootings. These shootings then become an explosive flashpoint of diverse interpretations, cautioning us to be wary of monolithic, biased, and self-serving explanations. Thus, I strongly propose that women's agency needs to be thought about in tandem with that of their targets and bystanders. Agency is always co-constitutive because the agency of the women stops or slows down where that of their targets begins or speeds up.

The contradictions that emerge from the shootings actually shape global misreadings of the women's act. The White House, human rights organizations, journalists, and local witnesses have all misread, unwittingly or not, the women's march as "peaceful" and the women as "unarmed." This erroneous reading has become one of the defining features of genital cursing and defiant disrobing when viewed in the postcolonial city and on the global stage. A day after the shootings, on March 4, 2011, Secretary of State Clinton issued a statement condemning the Gbagbo regime's acts of violence perpetrated against its people, "including his security forces' attack yesterday on *unarmed women* demonstrators that left seven dead" (2011, emphasis mine).

Five days later, on March 9, 2011, President Obama himself commented on the event:

> I strongly condemn the abhorrent violence against unarmed civilians in Côte d'Ivoire. I am particularly appalled by the indiscriminate killing of unarmed civilians during peaceful rallies, many of them women. . . . On March 8, the 100th anniversary of International Women's Day, we saw pictures of *women peacefully* rallying with signs that said, "Don't Shoot

Us," a strong testament to the bravery of women exercising their right of *peaceful assembly.* (emphasis mine)

The White House's rhetoric of "peaceful rallies" and "assembly" seems carefully crafted to avoid terms such as *demonstrations, marches,* and *genital cursing,* with their more aggressive and demanding tone. Further, the problematic narrative of "unarmed civilians" contradicts these women's accounts of the symbolic messages of their exposed nakedness and of their brooms and pestles. Genital cursing, whether used during the colonial period or in the postcolonial era, remains in some contexts the last resort against an unrestrained state power. The act of attempting to purify the land by causing the deaths of offenders is understood as heavy artillery within specific circumstances. Guns and exposed genitalia are both weapons, albeit differing ones. Whereas the power effects of guns are universalized and immediate, the effects of exposed and stigmatized genitalia accompanied by incantations are circumstantial and may be differentiated (a point I elaborate in scene 7). As such, I propose that we be more attentive to both women's own understanding of their agency and that of their targets. However, to be attentive does not necessarily ensure that such intended meanings would be shared, respected, or honored.

Obama and Clinton, intentionally or otherwise, are examples of a failure to correctly read the intentions of the protesters. The U.S. officials end up denying women the agency that their customary context bestows upon them. Failure to mention the mystical aspect of the "peaceful rallies" amounts to disavowing cultural prerogatives that the pressures of Westernization may not have wiped out. Clinton's and Obama's readings are akin to the shackles that have discursively constrained images of African women as helpless victims in need of saving. Naturally, most human actions are subject to multiple and often contradictory interpretations, but explanations of events offered by "leaders of the free world" carry actual political and cultural consequences.

The understanding that the women have limited abilities to predict the effects of their protests and rituals on targeted entities, or to shape the ways in which their actions are interpreted on the global stage, challenges us to be more nuanced in our conceptualization of their form of contestation. The reactions by women's male targets—dictators, public officials, neocolonialists, multinational oil companies—are important because they complicate the often one-dimensional accounts of the threatening removal of clothes as endlessly empowering for women. If anthropologists and historians have noted the cosmological meanings of genital cursing in societies where social cohesion was highly valued, the ingredients underpinning those valuations are slowly eroding in postcolonial contexts with the multiplying exceptional conditions that beget insurgent disrobing. The failure to gauge accurately the resolve of postcolonial bureaucrats to hold onto power and the slow disintegration of kinship ties, gender restructuring, and urbanization have exposed women to various forms of hostile counterattacks. These counterattacks assume many forms, including humiliation and slander, threats of arrests and actual arrests, physical brutality, and murder. Most of these attacks result from an emerging trend that I call the "secularization of defiant disrobing."

Toward an Affirmative Biopolitics?

As a concept and mode of reading, *naked agency* uncovers the slippages, lethal assumptions, murderous reactions, global misreadings, and conflicting explanations that undergird the threatening exhibition of mature female tabooed body parts in moments of sociopolitical duress. I relate the gesture and its complexities to the larger question of biopolitics in that it engages contexts determined primarily by the state of exception. The remaining part of this scene sketches the connection between naked agency and affirmative biopolitics, the specific growing subfield of biopolitical thinking that seeks to displace its negative strand. I call attention to the turn to the resilient aspects of life—being central to affirmative biopolitics—that refuse to be entirely subsumed under rhizomatic and seemingly sovereign forces. Specifically, here subjugation works hand in hand with countersubjugation.

Affirmative biopolitics is inflected in multiple ways by Roberto Esposito, Miguel Vatter, Rosi Braidotti, and Michael Hardt and Antonio Negri, making necessary a brief overview. One of the most important interventions in affirmative biopolitics is Roberto Esposito's (2013) call to think more about the politics of life rather than the politics *over* life.[7] The version of affirmative biopolitics he deploys moves beyond what he calls the "immunitary paradigm," which refers to the process in which life is shielded from that which gives it meaning. In other words, the protected life is eventually annihilated or emptied of meaning (Esposito 2013, 71).[8] Affirmative biopolitics *sensu* Esposito is named when the right balance is achieved between the two polar opposites of community and immunity, that which constantly seeks to protect itself from communitarian bonds.

For Miguel Vatter, natality and republicanism, as formulated in *The Republic of the Living* (2014), are generative for an affirmative conception of power within biopolitics, problematizing the separation between *zoe* (Agamben's naked life) and natality. Conventionally, these two concepts designate the biological sense of human life. However, Vatter, expanding Hannah Arendt's formulation, invests natality with an additional meaning: the foundation of human freedom, affirmation, and creativity grounded in biology. Vatter argues that an affirmative biopolitical practice ought to politicize zoe, not by transcending animality but by reconceiving zoe as a communal force of political formation. Conventionally, zoe—related to sexuality, family life, labor, and naked life, which are foundational to human power of creativity—is subjected to and yet excluded from biopolitical practices. In contrast, bios/normality relates to norms/nomos and indicates both negative and affirmative biopolitics of collective self-governance within the republic. In grounding natality as potentially emancipatory and as the foundation for a new "republic of the living," Vatter's conception aims to bring "an end to the government of human beings over each other" (5).

In "Posthuman Affirmative Politics" (2016), Rosi Braidotti, whose reflections share similarities with Vatter's, laments the rampant nature of negative biopolitics, writing: "This over-emphasis on mortality and perishability, which is characteristic of the forensic strand of contemporary social and cultural theory, reinserts the specter of human extinction and is haunted by the limitations of the project of western modernity" (39). She then advocates for an affirmative politics, deploying a posthuman paradigm that is part of the neomaterialism, or matter-realism, strand of thinking, as introduced by Jane Bennett's *Vibrant Matter* (2009). Neomaterialism is decidedly Spinozist and heavily committed to affirmation as a political energy and an ethical value. Proposing an account of subjectivity as "vital and self-organizing matter" (33), Braidotti argues for a posthuman definition of "Life" as zoe, which she defines as the vital force that connects all creatures independently of their rational capacities. This formulation is more inclusive as opposed to the binary thinking of "Life" as both zoe (life in the biological sense) and bios (the qualified form of life) that we inherited from Aristotle's political philosophy and that naturally includes some while excluding others from the possibility of a good life.

The work done by Michael Hardt and Antonio Negri shares similarities with these other conceptualization efforts. However, they emphasize embodied forms of contestation such as kiss-ins and sex strikes, modes of resistance that relate to naked protest because of their body-focused nature. Deeply committed to combining the Foucauldian account of power as a set of open strategies with the revolutionary possibilities in the Marxist analysis of capitalist production, Hardt and Negri (2004) mobilize "the multitude," a key figure in their political economy. They define "the multitude" as an alternative political organization of global flows and exchanges that can sustain empire as well as construct a counterempire.[9] With the multitude (multiplicity of singularities and a practice of power), Hardt and Negri reverse the usual framework by reconceptualizing labor and production to show how they can be seen as the locus of resistance in the biopolitical frame. Toward the end of *Multitude* (2004), in the "Democracy" section, a critique of traditional forms of political representation and a call for alternatives, they issue a pressing invitation for inventing new biopolitical weapons along the lines of Lysistrata's ancient Greek sex strike and Queer Nation's more recent kiss-ins. According to the authors, these weapons should be not only constituent but also capable of defeating the armies of the Empire for the construction of democracy.

Although Achille Mbembe in "Necropolitics" (2003) does not sketch a whole theory of resistant power, the end of his seminal article still brings to our attention the figure of "the suicide bomber." Engaging reflections by Elias Canetti, Martin Heidegger, and Paul Gilroy, Mbembe argues that the suicide bomber's very act of bodily duplication through death works to free her both literally and symbolically from the state of siege and occupation that operates in oppressive structures such as colonial occupation or the slave plantation. The meaning of such an agentive act is often repressed because it could not be accurately captured by current and dominant Western political categories.

In light of these reflections and provocations on a possible affirmative conception of power in biopolitics, naked agency, as a concept, draws on two major strands. The first is the attention to embodied forms of contestation, as outlined by Hardt and Negri (2004), and the second relates to Mbembe's (2003) reconceptualization of self-annihilation as agentive. The embodied forms of resistance brought to attention by the former include Queer Nation's kiss-ins, and sex strikes as imagined in Aristophanes's play *Lysistrata* and today echoed throughout the world from Japan to Togo and from Kenya to Colombia. First popularized by ACT UP (AIDS Coalition to Unleash Power), kiss-ins and acts of public disrobing have become fixtures of political actors whose main methods include direct action. Stripping naked or flashing sexual body parts are embodied oppositional tactics similar to the ones Hardt and Negri (2004) advocate, although analytics of gender and sexuality are absent in their theorizing. Further, attention to the lower social and economic class aspects that overwhelmingly define the women who resort to threatening self-exhibition in Africa is a supplement to the resistant cultural formation that Hardt and Negri name "the multitude."

However, while Hardt and Negri call for new weapons of protest, I note that specific segments of the African population, mature females, have marshalled for centuries what some today may consider new weapons. My work, unlike that of Hardt and Negri, does not highlight the absence of death in a vain search for seemingly sanitized weapons. While Hardt and Negri say, "We need to create weapons that are not merely destructive but are themselves forms of constituent power, weapons capable of constructing democracy" (2004, 347), I argue against the teleological-bound movement of sanitizing our theorizing of political militancy by moving from the destructive to the constitutive.[10] Based on the sociopolitical contexts examined here, if my exploration were to be shaped by these realities, the destructive cannot be separated from the constitutive, especially if destruction consists of purifying the land of those who trample its paramount values. Additionally, and most importantly, the rhetoric of triumphalism that Hardt and Negri (2004) marshal ignores the counterproductive effects of such embodied forms of political dissent, as I demonstrate with the lethal effects of defiant disrobing in several African contexts, specifically the March 3, 2011, all-women march. The paradoxical efficacy of naked agency belies such an overtly optimistic and teleological narrative that seeks to move from the destructive to the constitutive.

I am attentive to the inherently contradictory nature of human actions. And I rejoin Mbembe in acknowledging the presence of death in relations of power. In "Necropolitics," it is the martyr taking control of her death to free herself from the state of siege and occupation; in "naked agency," some of the women believe they can reverse the direction of death from themselves and toward the putatively protected, postcolonial bureaucrats. To assume that defiant disrobing is constitutive and not destructive is to subscribe to a sanitized version of social realities. As Julian Reid cogently observes, "Every regime of power entails its own particular account of who it is permitted to kill and how. An affirmative biopolitics is no different" (2013, 92). Similarly, in several African cosmogonies, naked self-exposure with its presumed ability to annihilate offenders is part of indigenous religious structures. The gesture is worked into the social arrangements as an instrument of conflict management.

However, to consider the customary meanings of naked cursing as outside history is to disregard the changing nature of human actions and their meanings. This is apparent in the simultaneous positions of victimhood and sovereignty that female protesters in the 1995 South African documentary *Uku Hamba 'Ze (To Walk Naked)* inhabit. With access only to the women's regrets and triumphs, I argue that the actors' agency is visible yet precarious because of their extreme emotions—fear, desperation, anger—and social and historical determinisms.

Dobsonville and the
Question of Autonomy

At the time when I was naked, I felt so powerful. . . . The power even to hit a police-man. —**Nthabiseng Hlongwane**, quoted in Uku Hamba 'Ze (To Walk Naked)

Men make their own history, but they do not make it as they please; they do not make it under self-selected circumstances, but under circumstances existing already, given and transmitted from the past. —**Karl Marx**, The Eighteenth Brumaire of Louis Bonaparte

Black-and-white archival footage of dozens of middle-aged women with ex-posed breasts and underwear opens the controversial 1995 documentary *Uku Hamba 'Ze (To Walk Naked)*. Directed by Jacqueline Maingard, Sheila Meintjes, and Heather Thompson, the film features women's resistance against the demo-lition of their shacks in Soweto, South Africa. Camera movement and distance significantly vary within the thirty-second tracking shot that oscillates between medium close-ups and medium long shots.

The vibrant and high-pitched chanting of "We want houses" combined with the slow motion of feet, hands, and exposed breasts seems like a mismatch of frames. The slow motion distorts bodies in a zombie-like fashion, which in-stantly captures the viewer's attention and focuses it on the women's breasts and overall body language. The images on the screen (slow motion) and the sound (high pitch and normal speed) seem not to belong to the same realm. At first, the disjunction is unsettling until the viewer realizes that the sound, although originally seeming unfaithful to the onscreen image, actually belongs with it. Further, the women's faces powerfully communicate their anger, although the camera remains distant. Perhaps the women's confrontational gestures prevented an extreme close-up.

At times, the viewer sees agitated naked women, bulldozers, dogs, shacks, and male onlookers. At other times, through the eye-level angle and the medium close-up shot, the viewer is assaulted by images of heavy-moving and exposed breasts as screaming women actively defy the camera and, by extension, the viewer. The women's body language evokes the *bolekaja*, a Yoruba term meaning "Come down and let's fight." This body language is not typically associated with (semi)naked black women's bodies. Images of sub-Saharan African women's exposed breasts and underwear should not shock anyone familiar with the Western representational politics that has dominated the visual landscape since the transatlantic slave trade. However, this opening scene can shock even the viewer who may have become desensitized to the neutralizing effect of excessive images of black (semi)nakedness. Through the naked protest of South African women against slum clearance, the black naked female body suddenly reveals more to us than we have expected. Unlike in racist colonial and neocolonial voyeuristic images, what is *not* in this documentary is passive nakedness. The women's self-exposure is active resistance against humiliation and dispossession. Their trancelike gestures could also suggest instinctive resistance with the intervention of blinding emotions, thereby curtailing the extent of their agency.

Given that often in instances of resistant disrobing, the viewer has access to exposed bodies with inaudible voices, *Uku Hamba 'Ze (To Walk Naked)* provides a significant opportunity to unpack the texture of naked agency. The documentary account suggests the intervention of empowering and possessing emotions in women's performance of nakedness, the women's internal sense of guilt and regret afterward, and the disapproving response of their community members, which all combine to demonstrate the fluctuating nature of the actors' positions of power and victimhood. Further, in light of the female naked insurgency during the late 1950s and the cross-ethnic naked shaming in the Fees Must Fall movement, the sense of naked protest in South Africa is markedly different from the sense we get from such protest in a West African context. This differentiation reveals the diversity of the continent as opposed to the faulty assumption of a unified, monolithic continent.

Uku Hamba 'Ze (To Walk Naked) Controversy

Five years after the women's gesture, the seven-minute documentary highlights the women's rationale before the event, their thinking or lack thereof during the act, and their feelings in the aftermath. The incident was initially broadcast nationally by the South African Broadcasting Corporation (SABC 2012), and the women's images were plastered across the front pages of *The Sowetan* and *The Star*, as well as other newspapers countrywide (Meintjes 2007). The specific form of protest—and not the forced removal of the protesters, a common occurrence in South Africa—caused the uproar and differing reactions, ranging from celebratory to scathing. What was absent in those reactions, however, was the voices of the women themselves.

Giving the women the opportunity to tell their own stories was a major goal for the filmmakers. The intentions of Meintjes, a member of the Women's National Coalition (WNC) Equality Research Supervisory Group (RSG), "were to ensure that the voices of all women would be heard and reflected in the new democratic order" (2007, 352). The idea of giving voice became more urgent in light of the lack of interest from journalists in allowing the women their voice: "Significantly, none of the women protesters had been interviewed by the press. Reporters simply recorded their actions. Even the World Television News journalists had not attempted to give voice to the women, except to visually record their naked protest for all the world to see" (353).

Despite its worthy goal, the documentary ignited some controversy about the power differential between the actors.[1] As white South Africans and academics, the filmmakers have a certain level of cultural capital, whereas the black South African women represented were mostly working class and living in poverty. Despite its attempts to signal the agency of the exposed bodies by departing from the colonizing (re)presentational politics that privileges the object nature of the Africans with an imposing narrative voice and immediacy, the film is still taxed with neocolonialism. The second aspect of the controversy concerns the politics of inflicting figurative injury on the exposed bodies, what Mieke Bal (1996) has, in a different context, called *double exposures*. The film figuratively inflicts the injury by further disseminating images of women's defiant nakedness.

The interest of "white" filmmakers in insurgent nakedness is a rising phenomenon. Several Caucasian British and American female filmmakers have released documentaries on defiant disrobing in Africa, including Candace Schermerhorn's *The Naked Option: A Last Resort* (2010) and Abigail Disney's *Pray the Devil Back to Hell* (2008). They revisit contemporary gendered traumas wherein women in Liberia and Nigeria stripped naked or threatened to do so to signal dispossession and vulnerability, to resist injustice, and to punish male targets.[2]

Power differentials and charges of exploiting African women's bodies through visual artifacts connect naked agency and female genital cutting (FGC).[3] The controversy surrounding *Uku Hamba 'Ze (To Walk Naked)* is similar to that for Alice Walker and Pratibha Parmar's documentary *Warrior Marks* (1993), which has been criticized for adopting a patronizing attitude toward supposed victims of a barbaric practice.[4] Discussion of female genital mutilation in the documentary makes, some say, the overt assumption that African societies deliberately butcher and mangle women en masse (Okome 2011). In addition, the topic of FGC is sensational and headline grabbing, impeding engagement in thoughtful scholarship. As I have suggested in the introduction, women in instances of naked agency and in female genital surgeries occupy liminal spaces, which curtails the simple dichotomy of victimhood and sovereignty.

Subtitled in English, *Uku Hamba 'Ze (To Walk Naked)* was released five years after the actual disrobing. The film uses black-and-white archival footage and full-color interviews in Zulu with three participants. The interviewees were longtime housing activists who mobilized various conventional strategies to make claims. Sadly, those strategies continually encountered the silence of housing authorities who were intent on demolishing the camp. When the women premeditatedly disrobed, some of them were arrested. Although the naked insurgency was useful in highlighting their predicament, it failed to secure the women housing because it took eighteen months of various claims-making strategies for the Transvaal Provincial Association to allocate them land elsewhere to build their own houses (Meintjes 2007). Out of the forty or so participants, only three—Nthabiseng Hlongwane, Thandeka Ndulula, and Thandi Mthaphe—were reportedly eager to reflect back on their uncivil disrobing. Interviews with the first two are filmed at medium close-up range to highlight the women's facial expressions during moments of anger and pain.

From Feeling to Inflicting Shame

In the South African housing protests, the women's goal in stripping to their underwear was to shame the police. These feelings were supposed to chase the police away and stop them from demolishing the shacks (Meintjes 2007, 348). In seeking to inflict shame, the women are liberally sharing their own feelings of humiliation. Given that defiant disrobing is their ultimate means of contestation, the women's backs are against the wall, having exhausted their affirmative options.[5] Resistant self-exposure here stands for the breaking of all taboos and is both a marker of vulnerability and a freedom from constraints.

More broadly, in combat, when one side finds nowhere else to go, that side paradoxically enters conceptually and experientially an empowering position, with both affirmative and negative options. They can continue to fight with an eye toward exhausting their targets' resources; they can peacefully surrender and acknowledge defeat; or they can surrender in a spectacular way that demands more resources and more efforts from their targets. These options are not exhaustive. The higher the number of women protesters and stakeholders, the higher the number of available options.

In this case, the protesters' expression of vulnerability is also one of sarcastic sexual availability. Ironic availability is evident in the women's chant, "We also have the 'thing' that other women give you! You can come and fuck me now!" (Meintjes 2007, 353). According to the filmmaker-scholar, this slogan refers to the corruption of officials whom the women accused of demanding sexual favors and money for housing.[6] Just as the revelation of tabooed body parts is meant to disarm opponents, the slogan is also meant to challenge by strengthening the version of virile masculinity that possesses women's bodies. Put differently, the mottoes of this event could be "strengthen to challenge" or "disarm by supporting." However, the women's veiled availability contrasts with the more defiant actual statement "We won't fuck for houses," reportedly one of the women's slogans and the subtitle of Meintjes's essay in a 2007 ethnographic companion to the film.

Through editing and montage, the documentary communicates the women's intense emotions of anger, shame, and fear. The interview with Thandi Mthaphe, one of the protest leaders who agreed to revisit the 1990s event, is visually framed by two shacks on either side of her standing body, communicating the physically assaultive aspect of the then-exposed body. Years after the incident, Mthaphe was able to tap into her initial anger, giving the interview a visceral force. She positions herself as a warrior with pumping fists. She beckons the "enemy" for a showdown as she exhorts other women to determination, defiance, and resistance, saying, "Strip off, strip off, strip off" at the same time that she says, "This shack of mine you shall never demolish. Today you have come to demolish my shack—you shall never." Regarding her sense of autonomy or lack thereof, she continues, "I wasn't even thinking that, by the way, I am a woman and that I am naked. . . . I also had a lot of strength and power but I was very hurt."

From her account of the exposing gesture emerges a supposed paradox and a contradiction. The supposed paradox concerns feelings of power and pain: "I had a lot of strength and power" and "I was very hurt." Although being hurt and feeling powerful are putatively opposed, the two states actually may engender each other. The contradiction in her confrontation has to do with her putative lack of rationality, "I wasn't even thinking," and an awareness of gendered constraints, "I wasn't even thinking that, by the way, I am a woman and that I am naked." How does one become aware of sociocultural constraints when one is not cognitively engaged? Although developments in affect studies and neuroscience have accentuated the cognitive aspect of emotions, it remains unclear whether becoming aware of one's situation is possible without cognition.

A second interviewee, Nthabiseng Hlongwane, strengthens a close correlation between self-exposure, typically thought to cause vulnerability, and feelings of power. The close-up of Hlongwane's face forcefully communicates her unmistakable sense of power in her facial expressions even though she was naked, as she says, "At the time when I was naked, I felt so powerful . . . the power even to hit a policeman" (figures 2.1 and 2.2). That a South African woman in her social position harbored that kind of (symbolic) power against a white police officer during the rule of the Afrikaaner National Government is remarkable. Hlongwane regretfully acknowledges that through her nakedness, she triumphed because the national outrage following the broadcast of the protest gave the women an advantage by drawing attention to the right to shelter that they so desperately lacked.

In exposing their nakedness, the militant resisters protesting forced removal sought to place their targets in a state of utter vulnerability, wherein they could be disarmed and confused, albeit temporarily. That moment is akin to, yet goes beyond, the Freudian "uncanny" for the male targets. Sadly, for the women and countless others like them, their expectations did not pan out according to their plans. In fact, the police reacted contrary to the women's expectations. Rather than feel shame and flee the scene, the police stood their ground and proceeded to demolish the shacks. Given the racial divide between the police and the women, the women's performance of defiant nakedness is a markedly volatile endeavor that perhaps has limits.

Determinisms

Although the protesting women reacted confrontationally, collectively, "unconsciously," and with a certain degree of autonomy (the gesture was reportedly premeditated), their sense of power and accomplishment was both short-lived and ambiguous. The intervention of emotions and various forms of determinism curtail the extent of the women's full autonomy and triumph. The limit to autonomy that social and historical determinisms inject is similar to what Karl Marx observed when he claimed that men do not make history as they please, which Lawrence Grossberg has recently formulated as "agency—the ability to

FIGURES 2.1 AND 2.2 Stills of Nthabiseng Hlongwane from *Uku Hamba 'Ze (To Walk Naked)* (1995). Courtesy of Jacqueline Maingard, Sheila Meintjes, and Heather Thompson.

make history as it were" (1993, 15).[7] Rejoining the Spivakian conceptualization of agency as institutionally validated action (Spivak 2009), I delimit the extent of women's untrammeled capacity to prosecute grievances.

Given the intense circumstances required to marshal defiant disrobing, it is unsurprising that stripping naked in public involves several emotions, the most important of which are desperation, shame, and anger. Such an expression of desperation and perhaps of instinctive resistance was powerfully articulated by Thandi Mthaphe, who justifies her action at the time of the demolition thus: "I had lost myself for a while because what was happening had disturbed me" and "I was heartbroken when my shack was gone, my shack, my home that I loved very much was gone." Here, Mthaphe perceives the demolition of her shack as her own demise. This feeling of dissolution and loss of self, of utter dispossession, and of psychic disturbance does not rely on outside reaction for validation. In her powerful analysis of resistance, Dina Al-Kassim refers to the women's sense of loss as them becoming "unintelligible to themselves and to their community" (2007, 126). Compounding her sense of loss and dispossession, Mthaphe adds, "I had forgotten that as a woman I wasn't supposed to walk naked among the people." Mthaphe's instinctive resistance is not exclusively instinctive because of the cognitive component of emotions. Her agency results from a certain awareness of her locatedness and from an understanding of her challenging material conditions, which stimulate her capacity for reaction in a specific way. The blinding and empowering emotions combine with determinist frameworks to limit the extent of the women's agency.

Social Scripting

The women's autonomy feeds on and yet weakens with the activation of prevailing social scripts. The women do not exclusively define these scripts, including the outrage around exposed genitalia, of which defiant disrobing is a constitutive element. These women therefore may be partly thought of as instruments of their exposed bodies, the vessels through which cultural meanings become communicated. During interviews for the movie, the filmmakers came across a possible exploitation of women's bodies. Meintjes reports in her essay that Benson Banda, a member of the Doornkop Civic (one of the grassroots organizations acting for equal housing rights), argued that one of the organization's strategies "was to get the women to take action, because it was felt that 'women are more listened to than men in the Commissioner's office'" (2007, 356). Whether Banda's allegation is founded or not is beyond the scope of this study. Suffice to say that, as we shall see in other scenes, women's insurgent self-exposure is not opposed to patriarchal stylistics of domination; actually, it works in tandem with them, as it can be mobilized to restrain that domination.

In addition to possibly having their nakedness instrumentalized, two of the three women who disrobed reported feeling defeated, haunted by the disapproval of their community and by having exposed their nakedness to the cameras. The last sequence of her interview and of the documentary features Mthaphe moving from physical exhortation to her feelings of defeat and regret after the demolition of her shack. Those feelings, later aided by societal disapproval, lead to her crying and saying, "I was very saddened because I had no home." The deeply touching close-up of Mthaphe's face in that moment resonates with the ambiguous nature of naked protest: anger, defiance, sadness, vulnerability, and regret coexisting in the same emotional and personal space. At the moment of their resistance, local community members expressed their bafflement at the women's loss of "discipline" with the injunction "Sit down! Hold on to discipline! Sit down!" Moreover, Thandeka Ndudula, the third interviewee, confessed that she tried to hide behind other protesters: "I was afraid of my husband seeing me." Clearly, the disapproval and incomprehension of community members show that the performance of nakedness is not always a communally authorized gesture, even in small communities. This constitutes what I call *social scripting*.

The presence of cameras that record, mediate, and preserve for posterity suggests that the naked women will have to answer for their public nakedness in future years. The film clearly captures Nthabiseng Hlongwane's sense of shame and regret.

> But now I have one problem when it comes to my grandkids because God has helped me find a place, now I become even more sad when I think of those things. I don't know who to blame, the government, the town clerk, the people I used to live with, or blame God. I don't know who I was because I was naked as anything. My body that is respectable at all times. Even now when I respect my body I ask myself, "Why should I respect it?" Because I'd shown it to the nations.

The act of exposing tabooed body parts that wrote these women into the modern history of resistance in South Africa during the country's transitional years of the 1990s, and the camera and technologies of preservation and dissemination that helped these women give visibility to their plight, continue to haunt them. Further, their embarrassment, produced by the disapproval of community members, testifies to the paradoxical nature of stripping naked in postcolonial/postapartheid settings. However, the different emotions are agentive, as they endow the women with the circumstances that impel progressive social and political change.

The protesters' sense of regret and shame is thought-provoking and even confusing, especially in light of Meintjes's report that it was Nthabiseng Hlongwane herself who came up with the idea of disrobing. Hlongwane reportedly drew on ideas and discussions she had heard from her grandmother, who raised her. If Hlongwane was in fact inspired by her grandmother's recollections and stories, did these suggest possible shame for the disrobed person? Alternatively, like many romanticization narratives of resistant nakedness, did the grandmother's stories leave out the counterproductive effects of the act? Perhaps the stories included those unexpected consequences, but desperation overtook the actor, blinding her to the possibility of shame. Or perhaps the disjunction between the era of the grandmother's stories—without the capturing and disseminating power of the camera, and without the transitional South Africa setting—accounts for Hlongwane's regrets.

With conflicting accounts of the origin of the act, the women's agency becomes more precarious. Some reports say that male activists suggested that the women disrobe, while some of the women suggested that they themselves came up with the idea. Both accounts support the argument that the self-exposure was a premeditated act. But if so, why would community members then chastise the women for their supposedly uncivil form of prosecuting perceived grievances and for their lack of respectable comportment? Here, we see multiplicity in community—that a community is not necessarily a unified entity, a point that Jean-Luc Nancy makes in *The Inoperative Community* (1991). Still other women protesters suggest that stripping naked was an unconscious act because their emotions overtook them. All the conflicting accounts of this instance of mature female insurgent self-exposure may be inherent to the gesture. They all contribute to making the defiant disrobing not one but, in fact, several acts. Moreover, to a greater degree than news reports, the documentary along with ethnographic accounts bring to the fore these contradictions, often induced by larger forms of determinisms, mainly historical.

Historical Determinism

A racist historical determinism would attribute the women's mode of protest to a pathological sexuality. That potential illegibility results from the resilient effects of colonialism that may reduce women's agency to their nakedness or their lack of rationality (more on this to come).

From the fifteenth century onward, the resilient afterlives of the historical injury inflicted on black bodies consisted in framing their (semi)naked images as animal-like and hypersexual. These images have consistently affected natural and social sciences and European voyeurism. Postcards and other visual artifacts of the colonial period, documentaries, and feature films—an archive that Johannes Fabian termed "soft pornography" (2000, 236)—have (re)presented (semi)nakedness as indexing savagery, sexual licentiousness, disorder, and immorality (Basden [1921] 1966; Roosevelt 1910; Weeks 1913).

One example of how this untrammeled sexuality was brought into being derives from the figure of Saartje (Sarah) Bartmann, also known as the Hottentot Venus, the South African woman who was exhibited in Europe as a freak. The supposedly disproportionate sizes of her buttocks and genitalia came to support the early nineteenth-century narrative of black degeneracy and sexual deviance. But, rebutting the tendency to generalize African women, novelist and cultural critic Zoë Wicomb frames Bartmann's fate and her overexposed genitalia as the symbolic representation of all colonized black women (1998, 91).

Through the erasure of individual identities, the trope undergirding most racist imperialist texts, independently of their locations and idiosyncrasies, is that black women "suffer" from pathological sexual drives. A number of sources document how the colonizing enterprise, mistaking the natives' absence of clothes for lack of modesty, forcefully "clothed" and often unclothed them as forms of brutal punishment and humiliation (Coly 2010; Comaroff and Comaroff 1991; Musisi 2002).

However, the colonialist framework has proven shaky. Numerous instances (e.g., in Cameroon, Kenya, Yorubaland, and Côte d'Ivoire) exist in which colonized women's self-exhibition indicates that they exceeded the colonialist paradigm that framed their bodies as pathological and irrational. This resonates with Rory Bester's observation that "from art to ethnography and back again, the unclothed black subject has consistently been the focus of voyeurism and science. But as much as such gazing has 'disarmed' women, so too have women used their nakedness to disarming effect, to claim or reclaim a power that is threatened or lost" (2003, 9).

The bodily manifestations, in front of television cameras, of the presence of spirits that these participants may be harnessing to expose desperation can be read as an index of the failure of the civilizing mission because it materializes the return of the repressed—that which the colonial enterprise sought to eradicate. The reading of the possibility of spirit possession highlights the question of human agency as it intersects with the agency of spirits, whereby the women demonstrate their inability to mobilize conventional forms of political rationality. In addition, one can consider as spirit possession the women of Dobsonville's sense of "loss," revealed in quotes from the documentary such as "I had lost myself for a while because what was happening had disturbed me."

Africanist literature in the social sciences and the humanities on spirit possession is vast.[8] Therein, spirits serve as indirect forms of commanding attention, rectifying grievances, and reaching concessions (Giles 1987). Because spirit possession is predicated on the assumption that the body and the self are not impervious entities but rather can be influenced, penetrated, and even taken over by an external spirit, it is presumably opposed to the rational mode of knowledge—thus, the colonial ideology's investment in erasing spirit possession alongside related practices.

The goal of colonialists in bringing "civilization" to the native in the form of Christianity and its corollaries, such as clothing the native and leading her to forsake her pagan gods, has been extensively discussed in African studies. During the Enlightenment, the religious, or rather the fetishistic, practices of Africans were framed as a sign of their fundamental otherness. In "Geographical Basis of World History," to pick a random example from a long series of possible racist texts of the Enlightenment, Georg Wilhelm Friedrich Hegel argues:

> Religion begins with the awareness that there is something higher than man. But even Herodotus called the Negroes sorcerers: now in *Sorcery* we have not the idea of a God, of a moral faith. It exhibits man as the highest power, regarding him as alone occupying a position of command over the power of Nature. (1894, 97–98)

To Hegel, the absence of a Supreme Being among Africans and the nature of their sorcery or fetishism is indicated by their attributing agency to objects that should have none: (1) The African recognizes the sun and the moon as powers in their own right, and claims that they have power over him. (2) But he does not see in them eternal value, and he can discard them at will. Hegel's confusing, while simplistic, explanation disseminated a commonplace misconception of the African as attributing agency to objects that ontologically lack it.

Using ritual speech among Marapu followers in Sumba and juxtaposing it with Dutch Calvinists and their account of the power of language, Webb Keane (1997) demonstrates that although a predictable dispute emerges about the locus of agency either in God or in divine subjects, ultimately both place the locus of agency (with some readjustment of scale) outside the subject. He then concludes, "Indeed, if we take seriously the ways in which the actions of subjects are mediated by objective forms like those of speech, we might ask whether *anyone's* practices could altogether escape some sort of charge of fetishism" (690).

Colonialists, however, did escape the charge of fetishism because they were invested in converting the natives to Christianity, with accounts like Hegel's subsequently used as an excuse for the colonization of the continent. Religious conversion of the native was of paramount importance in this process, so much so that missionaries preceded or accompanied explorers and other figures of the civilizing mission. Conversion is a form of discursive and material violence that maims the identity of the native.[9] Although colonization succeeded in South Africa (almost 80 percent declare themselves Christians), a government source, *South African Yearbook 2001/02* (Republic of South Africa: Government Communications), suggests the syncretic nature of Christianity in the country.[10]

Given that the "possession trance" of the women protesters suggests their temporary position as channels between the real and the spiritual worlds, the women may be conceptually imagined as lacking agency of their own. Thus, the possible trances speak to the precarious and unfinished nature of conversion, which, clothed in an erroneous and commonsensical European connotation, oversimplifies the murky process that involves syncretism and resistance and ignores a host of hybridizing selective appropriations. In light of the large number of Christians in South Africa and the sensationalism and outrage that the women's act begets, it is surprising that colonial ideologies in Africa did not foresee the resilience of "irrational" ways of being. These ideologies, rather than consisting of a coherent program with a rounded set of outcomes, were full of contradictory and half-finished currents and projects.

Making History

In conceptualizing new sociopolitical subjectivities in postapartheid South Africa, several scholars have pointed to, but not fully analyzed, the Dobsonville naked protest (Coombes 2003; Al-Kassim 2007; Meintjes 2007). Although not fully self-conscious model of revolutionary change, the women's embodied form of contestation introduces ruptures and reversals in what seems like an interrupted and uncontested march of postapartheid male-dominated and neoliberal values and practices. Rather than give up or give in, those who mobilize their nakedness, their last weapon, reveal an awareness of their conditions, their resourcefulness, and their hope for a better future, although the possibility exists that they will be defeated as agents, as was Draupadi, the defiant female protagonist of Mahasveta Devi's revisionist short story "Draupadi" that Gayatri Chakravorty Spivak (1981) famously translated and discussed. Gang raped and left naked with mangled breasts, Draupadi reclaims her unsightly, defiled, and humiliated naked body to defy and unsettle, albeit only briefly, seemingly unassailable male power.

Although such protests are not history-making in the sense of large-scale collective insurrections, or tied to the overthrow of systems of oppression, or even linked to ideologies of generalized emancipation, they do make history in a specific sense. In *Arguments within English Marxism* (1980), Perry Anderson relativizes and even challenges the unachievable Marxist notions of revolutionary praxis that dominated radical theories of change and agency during the 1970s. He identifies three categories of making history: first, the private sphere (getting married, for example); second, collective political militancy (military conflicts, political and social upheavals, religious movements); and third, the "unprecedented form of agency," which refers to global social transformations that have been considered to be history-making (Anderson 1980, 20). As social-science research has shown since the 1980s, the reduction of history-making to global social transformation that Anderson has criticized fails to provide a wide enough window through which to view the innumerable manifestations of sociopolitical militancy around the world.[11]

Further, the version of agency that is synonymous with individualism fails to account for the women protesters' action because of the collective nature of their organizing.[12] Often, collective actors do invoke a communal subjectivity in order to gain traction within liberal-based constructions of sovereignty that assume the individual as their epitome.[13] To demonstrate the parochialism of the account of agency that is understood as individualism, consider the Gilles Deleuze and Félix Guattari ([1987] 2007) critique of received concepts.[14] In the opening words of *A Thousand Plateaus*, they state:

> The two of us wrote Anti-Oedipus together. Since each of us was several, there was already quite a crowd. Here we have made use of everything that came within range, what was closest as well as farthest away. We have assigned clever pseudonyms to prevent recognition. Why have we kept own names? Out of habit, purely out of habit. To make ourselves unrecognizable in turn. To render imperceptible, not ourselves, but what makes us act, feel, and think. Also because it's nice to talk like everybody else, to say the sun rises, when everybody knows it's only a manner of speaking. To reach, not the point where one no longer says I, but the point where it is no longer of any importance whether one says I. We are no longer ourselves. Each will know his own. We have been aided, inspired, multiplied ([1987] 2007, 3).

Deleuze and Guattari's reflections are useful for the rethinking of the fantasy of the enduring, self-contained, and individualized agent. In linking the collectivism of women's agency to their anti-individualized agency, I seek to make the women's organizational scheme via Deleuze and Guattari's reflection more intelligible to audiences outside their primary epistemic settings. With the layering of viewpoints, experiences, and influences that all constitute an agent, gaining access to a self-determining autonomous individual among the women protesters may be difficult. Since the women's agency is always already multiplied, there is a negligible difference between each woman's individual agency and what can be called a "collectivist agent." In that sense, it is possible to read the South African women's act as agentive.

However, the internal tensions within the collective framework challenge the assumption of a unified communal endeavor where all members enjoy an equal share of contribution to the ultimate decision. Yet, unlike the dominant liberal model of agency, the version of agency that the women's insurgent nakedness embodies is collective, precarious, as well as triumphant.

Aggressive Disrobing in Recent South African History

With the use of nakedness in claims-making, the activists of Dobsonville stand at the junction of apartheid and postapartheid eras in South Africa. Before 1990, a few historical sources have pointed to self-exposure in protest, namely the 1959 beer hall boycott. After 1990, news channels reported on several events between 2001 and 2012. However, 2016 saw a mushrooming of naked shaming on university campuses. The rest of this scene is devoted to situating Dobsonville within the history of naked protest in South Africa and putting that history in conversation with discussions of ritual nakedness in West Africa.

The 1959 beer hall riots and boycott in South Africa stand as one of the few pre-1990 South African naked protests available in Europhone languages circulating internationally. In these events, the women mobilized conventional modes of contestation such as petitions, memoranda, and marches, but they also prayed, sang, clapped, chanted, and even stripped naked. The causes of the riots are to be found in a complex system of class antagonism with racial and cultural undertones. The apartheid system was to ensure the consistent supply of a well-controlled black labor force in urban areas, which was essential for the continued accumulation of capital for propertied classes. The 1908 Native Beer Act was the regime's calculated effort to limit the "corruption" of black minds and bodies, which the regime believed was caused by prostitution, domestic violence, and drunkenness. Restricting the natives' access to liquor and "kaffir beer," and eliminating local informal beer-brewing businesses, was thought to eliminate these vices. The act prohibited Africans from drinking beer except in government-sponsored beer halls. The 1920 Liquor Control Act and the 1923 Natives Act 21 also outlawed African home brewing for commercial reasons. These acts and subsequent laws affected women because they forbade women from visiting beer halls, limited their cultural prerogatives as beer brewers, endangered their household budgets as men spent their pay in beer halls, and, most importantly, attacked their economic sources of income, furthering female indebtedness and poverty. These acts prompted a series of women's antigovernment activities in 1929 although no historical records indicate the use of nakedness.[15] In 1959 the same reasons again spurred women's riots. Women not only protested the prohibition on beer brewing but also mounted an attack on the increasing coercion and social control by the repressive state. At the Rossburg beer hall on June 19, some women reportedly showed their naked buttocks to the police (Sadler 2002). Few details, such as reactions of the women's targets, exist on the disrobing, however.

Unlike the foundational 1929 women's public cursing in Igboland against the British colonial administration (Nigeria), which evoked scholarly reflections in several disciplines and in fiction, the 1959 South African insurgent nakedness received little scholarly attention in English and French. Subsequent events of naked agency in South Africa were recounted only in news reports in English- and French-language sources. In 2001 *The Sowetan* reported that in Kempton Park in the Gauteng province, eight women exposed themselves to protest against the police for denying them access to Bredell farm, where their belongings were being bulldozed along with their shacks (Fuphe 2001). Newspapers did not report similar incidents until five years later. In 2006 South Africa's IOL (Independent Online) reported that fifty female prisoners from the Mthatha prison, a maximum-security prison in the Eastern Cape, protested using their nakedness against their transfer into a different correctional facility in Queenstown, which would have placed the inmates far away from their families (IOL 2006a and 2006b). The transfer was to provide space for about one hundred male inmates. "It caused a lot of drama, because we were not expecting it," an official was quoted as saying (du Plessis 2006). Following their defiant disrobing, which attracted the attention of the media and disquieted prison officials but ultimately failed to halt the transfer, the women started a hunger strike. To this day, no one knows the result.

If no words, spoken or written, from the inmates and their exposed bodies made it into the public sphere, that was not the case in 2012, when the South African Broadcasting Corporation (SABC) incorporated a thirty-second newscast with the unequivocal title "Women Strip over Water Shortages." In the report, a crowd of men and women, young and old, hold makeshift placards, chanting and dancing. Some of the placards read, "Water and Road," "No water, no vote," "Pls Listen Us for Once. Our Concern Is Water!!" (figure 2.3). Several of the mature women in the crowd are naked, others are wearing white wigs, and all are speaking defiantly. The protesters' chants and words are muffled by cheers from the crowd in the background while the news anchor's voice dominates the soundscape. The police reportedly intervened when the women started undressing. While some of the women's exposed breasts are shown, other naked bodies are blurred but perceivable (figure 2.4). In addition to their resistant undressing, the women and other villagers erected barricades to disrupt government activities and services such as schooling in the village.

Despite the media attention to this event and others like it, scholars have not yet discussed these types of events in French and English. This absence of discussion contrasts with the already impressive amount of published literature on the 2016 cross-ethnic naked shaming on South African campuses (which I discuss in detail in scene 6, "Secularizing Genital Cursing and Rhetorical Backlash").

Scholarly studies of West African disruptive disrobing by women are much more available than are studies of the same phenomenon in South Africa. Occupying second place behind Nigeria's nine reports of collective female aggressive self-exposure, scholarship (in English and French) on South Africa is limited. Additionally, the sense of genital cursing as annihilating offenders and calling on ancestors to right perceived wrongs and to ward off evil is widespread in the scholarship on aggressive nakedness in West Africa. However, no mention of genital cursing is available in the scholarship about South Africa. This contrast is thought-provoking because most scholars refer either to seminal anthropological texts on West Africa or to their anecdotal evidence. I have yet to encounter any reputable study in French or English regarding South Africa that mentions how indigenous religious practices undergird the revelation of tabooed body parts as a means of cursing or shaming perpetrators for unacceptable behavior.

In her 2007 essay, Sheila Meintjes mentions the long-standing nature of nakedness in protest. But she cites Shirley Ardener's 1973 seminal study on West Africa as an elucidating source on South Africa. Annie Coombes also takes a comparable approach, writing, "There is a long tradition of women from many different African states resorting to naked protest in the face of official (usually

FIGURES 2.3 AND 2.4 Stills from "Women Strip over Water Shortages" (2012). Courtesy of South African Broadcasting Corporation (SABC).

male) reluctance to listen to their grievances and address their demands" (2003, 254). Both refer to the same foundational texts, as did Katherine Sadler (2002), all citing Judith Van Allen's 1972 essay on Nigerian Igbo women. If they do not reference the same sources, writers resort to anecdotal accounts to argue the customary aspect of the practice. Speaking of the prison inmates' aggressive nakedness, Carien du Plessis argues that the women "revived a unique tradition by staging a naked protest or *setshwetla*" (2006). Similarly, reflecting on the naked protest in Fees Must Fall, Mpho Mathebula also draws on anecdotes to explain the meanings of women's act.

> Growing up, when I sometimes visited a village in Mpumalanga, I would hear stories of old women who would walk in the streets naked. It was a form of protesting against social ills, bad judgements from the chief, any abomination in the village, or gross misconduct by the chief or the headman. Old women in the village would stage naked protests as a way of asking for rain from God and the ancestors during long, dry seasons. The old women would stage naked protests to nature and the gods of the land to give them rain. (2018)

In like manner, Teresa Barnes argues in a note, "The practice of naked protest by women in Africa is not unique to this situation [the 1990 women's uncivil disrobing]; even at the workshop people shared anecdotal evidence of it elsewhere. Full-body protests are, in one sense, made for postmodernist analysis, yet in another sense, it is such a difficult subject to represent that there have been no historical studies to date" (2002, 308). A 1991 conference/workshop in Cape Town was organized partly in response to the "Black Women by and for Ourselves" conference held at the University of Durban in January 1991.

Note that highlighting this widespread approach is not a call for published historico-ethnographic accounts to stand as authoritative explanatory frameworks. These sources, with their possible imperfections, would mildly suggest possible rationales for, and reactions to, the gesture. The scarcity, if not the absence, of ethnographic sources leaves available to postmodernist explanations some of the instances of disrobing that Barnes suggests. The differentiation in the cross-continental function of genital disrobing demonstrates that Africa must not be seen as a unified continent but rather as a heterogeneous continent with differing political histories and nation-states.

However, given the counterproductive effects of defiant disrobing, which I unpack in subsequent scenes, a postmodernist analysis, with its unstable framework, is closer than stable narratives to offering the most cogent account of the dynamic of victimhood and sovereignty that is inherent to resistant nakedness. Such a paradoxical dynamic through the prism of exploitation is operative in the next section of the book. In scene 3, "Africanizing Nakedness as (Self-) Instrumentalization," I engage historical accounts and newspaper articles from the late 1950s to the 2010s about Côte d'Ivoire, Cameroon, and Senegal to debunk the simplistic instrumentalization arguments that emphasize women as victims of powerful forces.

Section II

Co-optation

Africanizing Nakedness as
(Self-)Instrumentalization

The victory was small, yet beautiful, and *it deserved to be exploited*. (emphasis mine) —**Coffi Gadeau**, preface to Henriette Diabaté, *La marche des femmes sur Grand-Bassam*

These leaders are asking other people's mothers to strip naked before the minister. Why did they not ask their own mothers to strip naked? —**George Labeja**, mayor of Gulu in northern Uganda (quoted in Christina Okello, "Angry Women Strip Bare")

In 2015 the mayor of Gulu in northern Uganda, George Labeja, complained to international news reporters that before the 2016 presidential elections, some members of parliament were exploiting for political mileage the women protesting in land disputes. Whether the mayor's claim is accurate is beyond the reach of this book. What is generative, however, are the accusations of co-optation. These cases, which I call "Africanizing nakedness," are attempts by public officials to inflate the customary power effects of mature women's angry self-disrobing for factional political interests. The cultural capital of genital cursing, nominal or substantive, derives from the affective powers of customary practices and from the primacy of gender equality and "tradition" in international aid donors' specifications for funding. Simultaneously, female sociopolitical actors participate in that same dynamic of co-optation to instrumentalize genital cursing for personal profits. Indigenization and feminization are two results of the processes of co-optation and self-exploitation. Most events of defiant disrobing do not qualify as, or even pretend to be, genital cursing or shaming in the religious sense. However, by investing or saturating these acts with putative traditional meanings, these bureaucrats and other power brokers turn genital cursing/shaming into naked agency.

Co-optation is not a new phenomenon; it occurred throughout the struggles for decolonization. However, the numbers of these problematic strategies have not only persisted in the postcolonial era but also skyrocketed. Rather than being ephemeral, these developments represent an important postcolonial biopolitical form of negotiating power and authority and thus challenge us to look closely at the transactions between various stakeholders and customary women's organizations in order to gauge the specificities of the postcolonial nation-state and the place therein of female corporations.

The entanglements of women's secret societies in modernized African political institutions can be discovered in reading historiography and newspaper articles from the late 1950s to the 2010s on the Adjanou in Côte d'Ivoire, the Takumbeng in Cameroon, and the Usana in Senegal, despite the state's inability to secure legal or constitutional recognition for the women of these societies. Each society raises different questions and enables different insights, although the women's purification and cursing rituals are equally distributed between both ruling regimes and opposing voices. Some performances of genital cursing target opponents, while others serve seated governments, either to repress opponents or to participate in peace negotiations of their respective countries.

I claim that the dynamics of cooptation and (self-)instrumentalization both enhance and obfuscate women's agency. This process speaks to the women's ability to effect change given the importance of their modes of claims making in the sociopolitical configurations of African countries. The exploitation for economic gains and political mileage by others, however, obfuscates the women's ability to act and react because the exploiters undermine the women's sense of self-determination and self-representation, thus achieving little in advancing women's causes. This complex exercise moves beyond current and simplistic arguments of women as either sovereign agents or victims of powerful forces. These are some of the major theoretical stakes in considering indigenous female institutions and their rituals within postcolonial biopolitical circumstances.

The Adjanou

Of the three female institutions studied here, the Adjanou are the nationalized female indigenous sodality. In the popular imagination, the name *Adjanou* is sufficient to conjure a series of negative affects such as fear and intimidation. Thus, for decades, the Adjanou as a group have been used as national symbols and as political pawns. I attribute the long-standing entanglement of this "secret society" in colonial/postcolonial politics to several factors: its celebrated contribution to the birthing of the Ivorian nation; the publicized spiritual concerns of the first president, Félix Houphouët-Boigny; and the political and social vagaries of the postcolonial state. Since the colonial era, the society has typically followed the ruling political party, although it has at times moved into the opposition.

Popular Mythology

The long-standing imbrication of the Adjanou in both colonial and postcolonial political maneuvering is discussed popularly and academically. During the 2011 presidential elections, journalist Jacquelin Mintho (2011) goes so far as to attribute the defeat of the colonial regime partly to the Adjanou: "'The Adjanou,' the sacred and ritual dance of Baoulé women that helped Houphouët-Boigny during decolonization to defeat the colonial regime, is mobilized to purify, ward off, exorcise, and conjure evil spirits."[1] Since through their religious rituals, women birthed the nation with their son, the founding father, they are naturally recruited in the postcolonial era for exorcism and purification during social disruption, electoral campaigns, and national reconciliation efforts.

An exhaustive and authoritative account of the Adjanou is impossible. Thus, any readings grounded in anthropological and ethnographic texts will remain sketchy. The term *Adjanou* designates three interrelated entities: the indigenous cult of mature women; the women themselves; and their exorcism ritual, which is made of both public and secret components. The cult organized around women's bodies, which is "the most powerful fetish/entity in Baoulé culture" (Guerry and Chauveau 1970, 45). Childbearing and initiation maximize the vital energy and the social capital of the group. The centrality of motherhood that we have seen in other cultural contexts also operates in Baoulé society, explaining the constitution of elderly women's bodies as the source of mystical forces. According to N'Dri Thérèse Assié-Lumumba's work (1996), women aged fifty to eighty who have been initiated as members of the society are considered the ultimate protectors of their community, symbolizing the rebirth and continuation of the group. Their expected domain of intervention is to balance the cosmic energy and root out evil whenever it menaces the survival and prosperity of the land. In times of wars, pandemics, and droughts, Baoulé men enlist their wives to intervene via the invisible world with their seven-day purification, energy-enhancing, and exorcism ritual dance. The ritual is divided into two events: one public in which the women wear white and adorn their bodies with white clay; and one secret, held at night, wherein they reportedly dance in the nude. The secret performance is unequivocally forbidden to men, and, according to lore, a man's seeing it would result in his death.

In this cosmogony, women's participation in war is also mystical, with their mobilization and channeling of the earth energy, whereas men's contribution is both material and mystical. To end a war, to boost the warrior's morale, or to protect husbands and sons, women perform the Adjanou in two parts, which they accompany with a series of animal sacrifices to enlist ancestral forces. In her study, Assié-Lumumba (1996) powerfully specifies the meaning of the women's white clothing and clay. White is the color of mourning as well as a religious symbol, and the clay establishes a telepathic relation between the war front and the women in the village. A change of color reflects the development of events; for instance, clay turning red indicates that a man has lost his life, in which case female dancers have to redouble their efforts to send more energy and protection (Assié-Lumumba 1996, 98). It is in light of this belief system that during the struggle for decolonization, women offered to perform the Adjanou—in their collective mind's eye, the brutality of the colonial machine was the stark manifestation of cosmic disorder, the domination of evil forces. This mythology undergirds the participation of the group during the 2002 rebellion and the ensuing widespread violence.

Contemporary Practices

In 2002, at the outbreak of the armed rebellion in Côte d'Ivoire, several women began performing the Adjanou under the leadership of Nanan Kolia Tanoh, the female chief of the Baoulé village of Douakankro. The women were responding to the urgent call issued by Charles Blé Goudé, president of the Young Patriots, the radical wing of then-president Laurent Gbagbo's party that the rebellion sought to unseat. Goudé had exhorted Ivorians to resist rebel attacks against Côte d'Ivoire by all means necessary (Djidjé 2003). Evidently, these means included indigenous religious resources, such as a secret exorcism ritual by naked performers. This ritual includes genital cursing as a form of purification as the women invoke forces within their bodies to purify the lands of those who violate its normative codes. At the end of the society's seventh day of performance, rebel soldiers stormed the village, kidnapping and killing all the women except the sixty-five-year-old female chief, who escaped. To date, there are conflicting reports as to the number of female deaths, ranging from zero to eight. Some claimed that the massacre never happened.

The massacre of the Adjanou women created a debate between several political parties and their proxy journalists as each party attempted to co-opt the murders for overt factional political purposes. One camp included members of Gbagbo's party, the Front populaire ivoirien (FPI, Ivorian Popular Front). The other comprised newspaper columnists, such as Venance Konan, Nicole Bancouly Djinko, and M. A. Djidjé—from *Le Mandat*, *L'inter*, and *Fraternité Matin*, respectively—who supported the Democratic Party of Côte d'Ivoire (PDCI). (The PDCI has used the Adjanou exorcism ritual and exploited the Adjanou women in its rallies since the 1940s, the era of active anticolonial struggles.)

Many of the events involving the Adjanou discussed in this book relate to the public appearances of the women mainly dressed in white. These instances are more common in the postcolonial era and suggest the increasing appropriation of the Adjanou by political leaders. The public visibility of the dancers clothed in white markedly differs from colonial instances where the women were reportedly fully naked. It is safe to specify "reportedly" because given the secret nature of the naked ritual, very few outside accounts exist to corroborate the fact that the women were starkly naked. The return of the Adjanou to defiant disrobing (again from the perspective of the dancers) during the political crisis of the early twenty-first century (discussed in scene 6) represents a subsequent and further evolution of these uses of the Adjanou in Côte d'Ivoire.

Two months after the kidnapping and killing of the Adjanou women, *L'Inter* and *Fraternité Matin* covered the story because Simone Ehivet Gbagbo, the First Lady of Côte d'Ivoire, visited the Baoulé people to bear witness to their suffering. On March 3, 2003, Djinko chose the following hyperbolic title for her article in *L'Inter*: "'Un génocide des Baoulé est mis en oeuvre': Massacre des danseuses d'adjanou, L'unique rescapée témoigne" ("'A genocide of the Baoulé in the making': The massacre of the Adjanou dancers, the sole witness the massacre of Adjanou dancers speaks up"). In her article, the reporter chronicles the atrocities that the Baoulé reportedly suffered and quotes Tanoh as saying: "Responding to the call of the Young Patriots to resist the rebel attacks that divided Côte d'Ivoire into two on September 19, 2002, I asked the Adjanou women of my village to organize so we could exorcise the evil. We danced for seven nights. Following the mystical week, rebel soldiers stormed the village and kidnapped the dancers, who never came back" (Djinko 2003).[2] The report highlights the outrage of the Baoulé spokesperson, who complained that the world had remained silent on the genocide of his people. Singling out the dancers in the title is misleading for those who consider other alleged massacres as more horrifying than those of the dancers. However, given the importance that the secret society holds in the Baoulé imagination, their massacre carries stronger implications.

The day after the release of the *L'Inter* article (Djinko 2003), *Fraternité*

Matin, the oldest newspaper in Côte d'Ivoire and closely linked with the PDCI, published Tanoh's statements (Djidjé 2003). Four years after the massacre, in 2006, renowned Ivorian journalist and writer Venance Konan, himself a Baoulé, published *Nègreries, 1994–2006,* a book wherein he explained the massacre as the rebels' fear of being cursed by the mystical forces in women's bodies and their incantations. For the atrocities, he blamed the president of the Ivorian Parliament, Soro Guillaume, who was leader of the rebels at the time of the massacre.

Accusations about the massacre of the women continued to unfold, especially during presidential elections in both 2010 and 2015. In such politically charged moments, the Adjanou name carries political, social, and cultural capital.[3] The fact that only such moments stimulate references to the massacre speaks to the factional instrumentalization that is made of these women's lives or their deaths. The unintended consequence of such calculation is to drown out Tanoh's voice, often relegated to the space of the unhearable.

In preparation for the October 2010 elections, eight years after the event, a delegation of the then–ruling party FPI visited the village of the massacre, supposedly to recognize the wrongs done to the villagers, and to promise justice (*Notre Voie* 2010). Further, in 2011, Laurent Akoun, the FPI spokesperson, blamed the ex-rebels for the priestesses' massacre. In 2015 *L'Inter* ran a story in which a close collaborator of Charles Konan Banny, former prime minister and influential member of the PDCI who was running for president, threatened that Banny would expose the murder of the dancers by the rebel forces that helped current president Alassane Ouattara "seize" power in 2011 (Doumbia 2015). Days later, an article about the electoral campaigns analyzed the pros and cons of Ouattara's possible campaign managers (C. Kouassi 2015). The analysis suggested that the possible adversaries of Soro Guillaume threatened to recruit ten thousand Adjanou women in an election upset to remove him from office (C. Kouassi 2015). Whether the story is factual or not, the point is that the mention of the term *Adjanou* is powerful enough in the popular imagination to elicit fear of the women and concern for one's safety and that its employment for factional political mileage is a compelling strategy.

To most Ivorians, the recruitment of the Adjanou to counter rebel attacks, to oppose political adversaries, or to effect national reconciliation, which I explain later, is not an unconventional method of political practice. Indeed, the entanglement of the Adjanou with current political processes results from its appropriation in the 1940s by nationalist elites, who nationalized customary forms of organizing to frame the country's ability for self-determination against colonial rule. The imbrication was made possible because Félix Houphouët-

Boigny, the president at the time, was seen as a deeply spiritual person (Naipaul 1984). Practices such as those of the Adjanou were therefore channeled as the agency of the people in order for the nationalist elite to formulate its need for independence, a point that Partha Chatterjee has made in a parallel argument about the South Asian context (1993).

After this nominal political independence, the Adjanou did not move back to the grassroots level; instead, they became installed in the corridors of the modernized Ivorian public politics. From Ouattara's investiture in November 2011 after the contested October 2010 elections to Bédié's and Gbagbo's presidential campaigns, the Adjanou have either accompanied these leaders to the podium or purified the venues by performing their exorcism rituals during these massive official ceremonies. The group's influence also goes back decades. Germain Séhoué, an Ivorian journalist, published *Au nom d' Houphouet: Sacrifices au palais présidentiel* (1997), which placed the Adjanou in the presidential palace of Houphouët-Boigny in Yamoussoukro during his reign from 1960 to 1993. Reportedly, the Adjanou represented one of the three pillars of the president's power, the other two being the famous sacred crocodiles and the Djaa spirits (Naipaul 1984). Overall, however, it is impossible to authenticate all events narrated in Séhoué's book. He claimed, for instance, that because he feared for his safety, he was forced to omit multiple names; thus, the biography reads like a series of rather fantastical stories, with human sacrifices and crocodiles standing up and conversing with the president. The book also reads like an indictment of Houphouët-Boigny's reign, during which the leader supposedly indulged in occultism, ancestrality, and mysticism.[4] This presidency supposedly rejected rationality, with its twin promises of modernization and development, which the opposition had hoped to bring to Ivorians in the 1990s.

During the 1995 electoral campaign, hundreds of organizations representing all segments of the population, from trade guilds to youth groups, from gendered institutions to religious formations, were created to support the candidacy of the incumbent, Henri Konan Bédié. (Bédié had become the constitutionally elected interim president in 1993 after the death of the first Ivorian president, Houphouët-Boigny.) Henri Konan Bédié and Aliou Sibi, a political collaborator, proudly listed these organizations in the campaign propaganda document *Le pari de l'an 2000: Henri Konan Bédié, 35 ans au service de la Côte d'Ivoire* (*The Challenge of 2000: Henri Konan Bédié, 35 Years of Serving Côte d'Ivoire*) (Sibi and Bédié 1995), which contributed to the Bédié cult of personality. Unsurprisingly, given its decades-long entanglement with Bédié's party, PDCI, the Adjanou were represented as supporters by Le Mouvement de femmes Adjanou (MOFAY), the Movement of Adjanou Women (Sibi and Bédié

1995, 44). Whether MOFAY enjoys legitimacy among the Adjanou dancers or whether it is just a nominal creation for political instrumentalization cannot be determined here. But, the fact that Bédié's propaganda lists the secret society indicates his endorsement of the implication of the Adjanou in Ivorian political practices. That which was for decades framed as the national female institution of a religious nature was thus transformed into a factional movement. Amid boycotts and accusations of political maneuvering, Bédié was elected by 96 percent of the vote, with a turnout of 56 percent of the population.

The Adjanou continued their visibility in the public political sphere after the 1999 military coup d'état that ousted Bédié from power and installed Army General Robert Guéi. The transitional regime organized the October 2000 presidential elections and manipulated the process to eliminate all major candidates except for one underdog, Laurent Gbagbo. Following alleged electoral rigging, Guéi declared himself president, as did Gbagbo. These competing declarations unleashed a bloody struggle for power by parties that were denied participation in the elections, leading to a war that caused atrocities and hundreds of deaths, including a mass grave of fifty-six young men. Following these events, Gbagbo emerged as president of Côte d'Ivoire, Guéi was dead, and Alassane Ouattara of the Rassemblement des Républicains (RDR, Rally of the Republicans) and Bédié of PDCI remained as the opposition. In 2002 the Adjanou continued their allegiance to the party's political campaigns and rallies, so much so that an April 2002 issue of *La Voie*, the newspaper closely associated with Gbagbo's party, published "Côte d'Ivoire: Coulisses du congrès: L'Adjanou, danse officielle du congrès" ("Côte d'Ivoire: Behind the scenes of the convention: The Adjanou, official dance of the convention") (B.s. and G.g. 2002; in Côte d'Ivoire, it is common for journalists to use pseudonyms or acronyms for fear of political persecution). The reporters castigate PDCI's past failures and its politicization of the Adjanou. A few months after this criticism, as mentioned earlier, the Adjanou mobilized to support the Gbagbo regime against the September 2002 armed rebellion that threatened its power.

In 2010, eight years after the Adjanou first opposed, then supported, the ruling regime of Gbagbo, the society once again opposed Gbagbo by demonstrating with other organizations against what opponents denounced as Gbagbo's warmongering, antidemocratic, and anticonstitutional practices. The president's unilateral dissolution of both the government-issued peace accords and the National Electoral Commission led to a series of protest demonstrations and exorcism rituals with genital cursing. The residents of Didiévi, in the center of Côte d'Ivoire, actively participated in the upheaval; an undetermined number of Adjanou performers walked to the prefecture, causing the prefect to flee. It is unclear from news reports whether the prefect fled in fear of women's mystical forces or from his fear of being physically assaulted. However, in other cities, the Adjanou failed to "defeat" Gbagbo's supporters because security forces reportedly beat their dancers (Loukou 2011), demonstrating how postcolonial circumstances put pressure on the mystical nature of women's rituals. In November 2010, after the declared victory of the PDCI and allied political parties during the first round, the winners recruited the Adjanou to perform. This performance, however, did not involve ritual nudity but rather the more palatable symbols of the society such as white clothing and white clay designs on bodies and faces.[5]

The second round of elections in November failed to bring political stability, and the Adjanou continued to be visible. Gbagbo's refusal to relinquish power created a political stalemate, even though the Commission Electorale Indépendante (CEI, Independent Electoral Commission) and United Nations observers declared the challenger, Alassane Ouattara, winner of the elections. In reaction, an armed rebellion sought to unseat Gbagbo while rumors of disappearances of members of the opposition, allegedly caused by the death squads of Gbagbo's camp, plunged the country into chaos. It was in this atmosphere of perishability that in February 2011 dozens of reported protest demonstrations around the country included the Adjanou. Following the war that the Adjanou could not prevent, the group was again enlisted in national reconciliation efforts.

Peace Negotiations

The August 2012 National Reconciliation event in Bouaké, the second-largest city in the country and the site of extreme forms of violence during the decade-long period of sociopolitical instability, was a first step toward reconciliation. After speeches by political and administrative leaders, prayers by leaders of non-indigenous religious groups (Muslim, Catholic, and Protestant), and exorcism by traditional leaders, hundreds of Adjanou dancers sprinkled the venue with potions and clay powder for purification purposes. The national president of the Commission Dialogue, Vérité et Réconciliation (CDVR, Committee for Dialogue, Truth, and Reconciliation), Charles Konan Banny—who had held various high-profile positions in Côte d'Ivoire, including that of prime minister from 2005 to 2007—presided over the ceremony. Banny was escorted to the podium by two Adjanou dancers (figure 3.1). He reminded attendees of the participatory aspect of reconciliation and peace, warning that none should take lightly the implications of the symbolic and spiritual aspect of the purificatory ceremony, for "one must not play with the spirits and one must not trick God" (A. Kouassi 2012).[6]

Cérémonie de purification et de deuil des violences par le CDVR à Bouaké / Charles Konan Banny : « Que tous les fils de la Côte d'Ivoire renouent la voie du développement de la paix et du dialogue »

Publié le vendredi 31 aout 2012 | Ivoire-Presse

FIGURE 3.1 Ceremony of Reconciliation (2012). Côte d'Ivoire political leader Charles Konan Banny in Bouaké with Adjanou dancers. Courtesy of *Ivoire-Presse.*

Although encouraging for peace building, the apparently seamless mobilization of multiple and often conflicting religious denominations by organizers of the ceremony seems more like a cosmetic arrangement than a genuine religious cohabitation. How do all these orders coexist without acknowledging the inherent contradictions that are at the core of their differences? Some Abrahamic religious leaders still consider indigenous religious practices such as the Adjanou purification rituals to be devil worship. And the differential distribution of state and constitutional recognition that Makau Mutua (2002) and Rosalind Hackett (1996; Hackett et al. 2014) argue crystallizes with the moral injury that the Adjanou complained about in February 2011, as I discuss in scene 6 in the context of deritualizing religious practices as rhetorical backlash. The group's involvement in the background and foreground of political calculations in Côte d'Ivoire is a differential repetition of the colonial-era practice of the Adjanou.

Colonial Practices

Several historical studies and documents of the PDCI-RDA, the major political party in Côte d'Ivoire for more than three decades, show that the party first nationalized and then overrated the Adjanou's contributions to the successes of the PDCI-RDA against the colonial administration (Amondji 1984; Assié-Lumumba 1996; Coquery-Vidrovitch 1997). The society was introduced into the 1949 political upheaval by N'Doli Amoin Madeleine, the first woman to perform the Adjanou ritual in the anticolonial context. Referring to the dance itself, N'Doli says, "Of all the female protest events, the most significant one was the Adjanou. The most important and the most serious is the Adjanou" (quoted in Ministère de l'éducation nationale et de la recherche scientifique 1985, 149).[7] She explained that among the Baoulé, women perform the Adjanou to strengthen their men and to give them the power to accomplish exploits, such as wage wars. It is in similar circumstances in February 1949, when the French colonial administration incarcerated Ivorian nationalists for subversive practices, that N'Doli reported, "I heard my Adjanou. I put on my 'kodjo' [menstrual cloth] and we performed the Adjanou throughout the city to bring good fortunes to our men, to help them succeed. Since we did not have guns, we thought that we needed a big number so that the whites understand that Houphouët was heavily supported by his people" (quoted in Ministère de l'éducation nationale et de la recherche scientifique 1985, 149).[8] From her account, the Adjanou functioned in two ways—mystical and material—to empower men and intimidate the colonial administration by communicating a sense of popular support for the nationalist cause. However, cracks in the narrative regarding the perceived public awe of the Adjanou and their genital cursing are already apparent in the speaker's counting on the number of participants for a show of force.

Following N'Doli's Adjanou ceremony, Marcelline Sibo, the leader of Agni female supporters, performed another Adjanou ceremony with her followers. She was arrested on August 4, 1949, by colonial police, who reported that an individual had complained to colonial authorities about women and their nocturnal noises. These noises were caused by several groups of women who danced the Adjanou throughout the night, rotating every two to three hours. (Note that the complaint dismissed the gravity of the women's undertaking, framing it as "noise.") The charges against Sibo included "death threats by fetish" (Dadié 1950, 79), night disturbance, public offense, and organization of a public event without authorization. Women demonstrated, and even walked twenty-five miles to Grand-Bassam in order to obtain Sibo's liberation. In December 1949, to protest the arrest of local political leaders, the women organized a more elaborate demonstration. Sympathetic journalists stated the number of female protesters as about four thousand, whereas historian Henriette Diabaté (no relation to the author) suggested that the number was closer to two thousand (1975). Diabaté also refrained from exaggerating the women's achievements, reporting that their direct actions failed to free the prisoners.

Although women in general initially offered their support primarily to party leaders, eventually they would be called upon when men's mobilization was thought to be either ineffective or impossible. Ouegnin Georgette, one of the women's leaders, highlights the party's reliance on women, in her role as the liaison between the women and the party leadership (Touzard 1987d). For example, the women preempted Houphouët-Boigny's arrest by barricading the bridge between the two cities. Another women's leader, Marguerite Sacoum, reiterates the indefatigable support that women provided to the party thanks to their organizational capacities (Touzard 1987d).

However, for political propaganda purposes, the elite essentially embellished and coopted women's mobilization. For instance, newspaper reports in France hailed the mobilization of women as the most spectacular female political movement against colonialism in West Africa. In his preface to Henriette Diabaté's historical study of women's protest demonstrations, *La marche des femmes* (1975), Ivorian statesman and former minister of justice Germain Coffi Gadeau thus describes the overt exploitation that was made of women's actions: "The [women's] victory was small, but so beautiful, *and it deserved to be exploited*" (Gadeau 1975, 3, emphasis mine).[9] Perhaps the news of the 1929 Igbo Women's War, or of the 1947 demonstrations by women, spearheaded by Fumilayo Ransome-Kuti in what is known today as Nigeria, had not reached those who celebrated the exceptionalism of Ivorian women. The reference to these Nigerian examples (1929 and 1947 among others) should not be understood as downplaying the women's courage. Rather, I highlight the ways in which women's involvement has been exploited for reasons divorced from a genuine recognition of their impact.

The exploitation of women's ability to mobilize massively against colonial governance is evident in several PDCI-RDA documents from the 1980s to the first decade of the twenty-first century, including Philippe Touzard's 1987 series *Mémorial de la Côte-d'Ivoire* (4 vols.), the Ministry of National Education's *Education civique et morale* (1985), and others, which continue to reference women's impact and sacred dance. In 1976 the one-party state honored women's courage with a picture of Marie Koré, a staunch female nationalist who also danced the Bété women's exorcism ritual against the colonial administration, on the 1,000 CFA banknote (1976–81). (Bété designates one of the four major ethnic groups in the country.) In 1999 a statue was unveiled in Grand-Bassam to celebrate women's contributions (figures 3.2 and 3.3). The accompanying inscription reads, "To our brave women, who by their historic march of December 24, 1949, on the Grand-Bassam prison, took back the freedom denied to our men."[10] Despite this aggrandizing rhetoric, very few documents exist about the statue. In June 2016, when I visited the Musée des Civilisations de Côte d'Ivoire, I could not locate any substantive information about the statue.

Public officials have sought to naturalize women's support in the party, demonstrating an expected alliance between them. Attributing the birth of the Ivorian nation to women and their spiritual and material forces strengthens the idea that women cannot disavow their own fruit, the party-cum-state-cum-nation (Touzard 1987d). In a way, the narrative works to control women's political participation by perpetuating gendered factional alliances. Yet, although women birthed the nation, according to this rhetoric of exploitation, that nation has actually seen little change in gender equality in modern terms, except for token ministerial cabinets of education, family, health, welfare, and women's affairs. As happens elsewhere around the world, even in countries where women were not as heavily credited as having birthed the nation, these specific cabinets remain their domains (Waylen 1996). In Côte d'Ivoire, different regimes have put little effort into meeting gender-based demands although founding leaders "exploited" women's movements.

The use of the Adjanou label by party leaders to designate all female modes of participation homogenizes a plurality of ethnic groups, thereby silencing ethnic differences and the ways in which different ethnic groups may have drawn from their cultural specificities to counter colonial administration. For instance, explaining how different groups organized to dance and relay information to each other, Diabaté (1975) quotes Georges Grah, who said that Marie Koré, a headstrong woman from the Bété ethnic group, mobilized the Bété women to perform the Adjanou according to their own culture.

To oppose colonizers, nationalist elites had to present a unified front in terms of gender, class, and ethnicity. For class specificities, Diabaté (1975) shows how only an elite class of women participated in the decolonization struggle. I propose that Koré's example demonstrates that the ethno-politicization of female organizing, which was already perceptible in the colonial era, widened during the 2010 electoral processes. Against the nationalization of the Adjanou, the recent appearance of another female political movement, the Kodjo Rouge in the public sphere both in Côte d'Ivoire and in France (scene 7) to support the regime of Laurent Gbagbo, indicates the heterogeneity of indigenous forms of contestation, thereby exposing the nationalist fantasy of national homogeneity.

FIGURES 3.2 AND 3.3
Statue and inscription
commemorating the
December 1949
Women's March on
Grand-Bassam prison.
Photos by the author,
2018.

A NOS VAILLANTES FEMMES QUI, PAR LEUR MARCHE
HISTORIQUE SUR LA PRISON DE GRAND BASSAM
LE 24-12-1949, ONT ARRACHE LA LIBERTE
CONFISQUEE DES HOMMES.

D. 2-1999
Mairie F. ABLE

Despite exploiting women and their rituals, the country has seen little gender equality. Women continue to lag behind men in terms of political representation. Côte d'Ivoire ranked 133 out of 144 countries according to the 2017 *Global Gender Gap Report* by the World Economic Forum. The 11 percent of female parliamentarians are far below the regional average of 21 percent. In November 2012, the minister of women's affairs introduced revisions to the marriage law that was to give equal rights to men and women. But PDCI members of Parliament blocked the bill, leading to the surprising dissolution of the government by Ouattara. The statute was eventually passed after the appointment of a new prime minister.

It would seem that the purpose of the introduction of a gender equality law would be to redress historical injustices behind women's economic and political disenfranchisement. However, this marriage law was merely another attempt to use the rhetoric of women's empowerment for political maneuverings around the country's economic growth. As it became clear, the law was a means for the country to qualify for nearly US$1 billion in development funds from the U.S. Millennium Challenge Corp. (MCC), a Washington-based government aid program (Bax 2013). These kinds of exploitative calculations by postcolonial bureaucrats are possible because of the preeminence of categories of gender and tradition as criteria for funding projects.

The Takumbeng

In its relatively brief occupation of the public sphere in Cameroon, the Takumbeng female society of the country's Northwest Region came to the fore in the 1990s with its genital cursing that targeted the ruling regime of Paul Biya. Similar to other indigenous female structures discussed in this book, the Takumbeng group is made up of postmenopausal women whose bodies are thought to possess mystical entities, which, when collectively mobilized and exposed, may heal the community of drought, epidemics, wars, and perverse despotism.

The society provided indefatigable support to Biya's opponent, John Fru Ndi, whose mother was the society's leader at the time. The fervent moments of popular political agitations were afforded by "Africa's Springtime," which was inspired by global conditions, including the 1989 fall of the Berlin Wall and the famous La Baule speech in 1990 by French president François Mitterrand (Bourgi and Casteran 1991).[11] This democratic awakening caused the fall of thirty-five regimes on the African continent, but regimes such as Biya's remained standing despite the pressures of the Takumbeng, popular unrest, ghost-town campaigns, and economic boycott.

The indigenous female sociopolitical institution that appeared in the public space was an adapted version of the organization of precolonial and colonial eras. Not only did the group intervene in nonspiritual ways such as distributing tracts, but it also reunited urbanites and women from multiple ethnic groups. Involvement of the Takumbeng in 1991 through 1993 lasted from the electoral campaign to the period of postelectoral violence and the declaration of a state of emergency. Their actions included handing out tracts to security and defense forces; monitoring ghost-town campaigns; barricading the residence of John Fru Ndi, the main opposition leader who declared himself winner of the October 1992 presidential elections against Biya; blocking the advance of the military; and preventing Fru Ndi's arrest by stripping naked as a genital-cursing gesture. The existing literature on the Takumbeng (Awasom 2003) explains its participation in the 1990s wave of political effervescence as accomplishing the sociocultural tasks of alleviating the suffering that Biya's dictatorship imposed on the group's children.

This rhetoric of naturalized duties is not a satisfactory causal explanation. I argue that the postcolonialized Takumbeng that had as its leader Fru Ndi's mother, "a woman full of determination and ready to die for her son" (Awasom 2003, 410), predominantly accounts for the defiance that the "traditional" female institution demonstrated in Cameroon. The Takumbeng highlights two dynamics, co-optation and self-instrumentalization. Fru Ndi's mother's role is one of interested leadership, which resembles the exploitation of institutions for factional rather than communal interests. However, given the nature of the Cameroonian nation-state, what counts as factional rather than communal is already problematic. Additionally, the society may practice self-instrumentalization by its clear understanding of and participation in the exploitative tendencies at play.

The Usana

Between the Adjanou, who are co-opted, and the Takumbeng, who self-instrumentalize, the Usana of Senegal exemplify both dynamics.[12] Although the Usana's first nationally recognized defiant self-exposure event concerns the student strike of 1980, the nominal or substantive imbrication of the society in the Casamance peace process started in the early twenty-first century. The Casamance conflict began in 1982 with the on/off low-intensity rebellion, allied with a political branch, the Mouvement des Forces Démocratiques de la Casamance (MFDC, Movement of Democratic Forces for Casamance), and pastored for years by Catholic priest Father Diamacoune Senghor. Originally, representatives of the insurgency sought the independence of the Casamance region, which is ethnically and religiously distinct from the rest of Senegal. Located in southern Senegal, Casamance is home to the Jola ethnic group, many of whom are animists and Christians, unlike the rest of Senegal, which has a Muslim majority. Since its beginning, the conflict has destroyed hundreds of villages and compromised the local economy, with its production of rice, vegetables, and fruit, as well as its tourism.

Named after the baobab tree in Diola under which the first society was reportedly created, *Usana* is an umbrella term that designates associations of urbanized Diola women who are pastored by females with mystical abilities (Foucher 2007). These are women whose potency, according to their customary episteme, resides in their ability to purify the community, heal social woes such as drought and pandemics, and punish violators of normative codes by casting a spell on them. The first of these urbanized female ritual societies was reportedly created in the 1930s–40s; by the 1990s, dozens existed. Their visible occupation of the public space takes the form of propitiation and purification rituals. The urbanization of these indigenous female associations responds to the women's crucial need to maintain some form of contact with their villages and their problem-solving networks because most powerful lineages and families have their own "fetishes" to which they resort in challenging moments (Foucher 2007). Successful imbrication of the rhetoric of "tradition" into Senegalese postindependence life and governance was encouraged by Léopold Sédar Senghor, the first Senegalese president and one of the founding fathers of the Negritude movement. Even currently, high-ranking politicians court the Usana for their blessings because of the cultural and political capital they enjoy.

The January 1980 Djignabo High School protest in Ziguinchor, which was part of that year's nationwide Senegalese sociopolitical unrest, resulted in the death of one student by police shooting. In reaction, thousands of women, under the guidance of the Usana, donned their traditional attire, such as calabash hats, branches of green leaves, necklaces of colored beads around the chest and the belly, and red brooms. Reportedly, they even exposed their tabooed body parts while chanting incantations and prayers to the deities. The women visited the governor's office, the police station, and the jail, demanding the release of students, which officials granted the next day. The fact that the Usana and student demands were granted popularized the general conclusion that the Usana hold specific kinds of powers to which politicians concede. Further, according to Vincent Foucher (2007), shortly after the female exorcism ritual, rumors circulated about the death of the governor's or the headmaster's daughter, strengthening the belief in Usana's magical powers. Problematically though, Senegalese statesman and university professor Assane Seck (2005), in his study of democratic emergence in Senegal, qualifies the Usana's 1980 genital flashing ritual as pacific, as other scholars have done in other contexts, a characterization that I question. Specifically, the qualifier "pacific" is used consistently in these studies without definition; authors assume the self-explanatory nature of the term.

Since 2000, the Usana have increased their visibility in the public sphere, acting as a third-party facilitator in the conflict opposing the government of Senegal to Casamance separatists. With countless failed ceasefires, decades of negotiations, and innumerable unexploded land mines, the urgency of the conflict has mobilized multiple third parties, including the Usana. The Usana attribute their involvement to multiple and promising local, national, and foreign circumstances, including the cultural capital of tradition, the investment of aid donors in gender equality, the rise of Track II diplomacy, the acute military fragility of the state, conflict ripeness, and political clientelism.[13] Initially, however, the Usana suffered from the wrath of the Senegalese government when the army destroyed their place of worship that was forbidden to men, a prohibition that created terror in nonmembers (Foucher 2007). The intervention of the army demonstrates the lack of constitutional and legal protection that imperial religions such as Islam and Christianity routinely enjoy. Despite the clause in the Senegalese constitution protecting freedom of religion and despite the social capital of tradition, religious expressions and practices, including those of the Usana, still do not benefit from the state and juridical protections to which the society should have recourse.

Eventually, members of the civil societies, backed by international aid donors, drew on their political organizing, such as union and community activism, to highlight the necessity of including women's skills, status in the public consciousness, and modes of effecting change in the peace process. Formalization of the involvement of women's groups into the conflict resolution became reality in December 1999, when the Forum of Women for Peace met in The Gambia with women from diverse groups (intellectuals, community activists, and "women of the fetish"—the Usana) (Foucher 2007, 68). These associations contributed to conflict negotiations by attending meetings with the government and members of the MFDC, calling for peace, and organizing ritual ceremonies, often to free separatists from the spiritual oaths they took and that bound them to the war (Foucher 2007). These ritual ceremonies suggest the widespread belief in the existence of invisible forces and their impact on the visible world. More importantly, as Stephen Ellis and Gerrie ter Haar (2004) have argued in other contexts, these rituals are predicated on the belief that evil or impurity has colonized the land and caused conflicts, and that it is therefore vitally necessary to root out the evil. Sharing in this view, international donors even funded sacrifices honoring the Usana and other fetishes (Foucher 2007). These forms of visible occupation of the public space fit squarely in the literature on Track II diplomacy processes thanks to their unofficial, secretive, and diffuse nature. Moreover, the mobilization of tradition, nominal or substantial, by international donors also creates the conditions of possibility for cross-instrumentalization by the women themselves, by local politicians, and by local development brokers with their own interests.

As already argued, gender terminology and objectives have become a condition for funding by donor agencies. Aware of those international funding mechanisms, local actors of many stripes use one another for their personal gains. Reading Vincent Foucher's (2007) reflections on the Usana of Senegal and Beniamina Lico's (2014) study of a local development broker is useful in uncovering the multiple and crisscrossed interactions between the interested actors. Foucher and Lico observe that semiprofessional and professional female facilitators and community activists understand the expectations and norms of the development system, or the "rescue industry," as Laura María Agustín (2010) has called it in a different sociopolitical context. Through their familiarity with both local practices and international funding agencies, these actors "recruit" or mobilize the Usana, becoming mediators and translators with donors.

Specifically, some of the Usana representatives are advised on what to wear and what to say in order to authenticate their belonging to an invented tradition that is packaged and sold to meet the objectives and language of international development agencies but does not match the quotidian practices of the women (Lico 2014). The requirement of authenticity is part of a larger neoliberal agenda that Elizabeth Povinelli demonstrates in the case of the Aborigines of Australia in her groundbreaking study, *The Cunning of Recognition* (2002). The women's political practical understanding allows them to move seamlessly within the corridors of the development industry for the negotiation of a space of power among three exclusive systems of authority: the postcolonial nation-state, indigenous religious practices, and international aid donors. For instance, C.B., the development broker in Lico's study, is reported as saying that she worked in "collective management of women's agricultural activities," "microcredit loans," and various activities with the Usana in "peace building" operations (Lico 2014, 136). In those capacities, local development brokers sell their knowledge as project managers for nongovernmental organizations (NGOs) and international donor agencies (Foucher 2007).

In addition to modes of self-presentation and fashioning consonant with international development standards of authenticity, these mediators and translators also manufacture, at times, a unified traditional Usana. For example, when ideological differences between the Usana emerged shortly after their initial recruitment, the facilitators went to work. Some Usana preferred to focus on the mystical purification of diseases, refusing to participate directly in political organizing for peace building; others held a staunchly political and separatist stand in line with the MFDC. The local brokers recruited the national tragedy of the Joola ferry sinking in September 2002 that killed 1,863, connecting the conflict with mystical causes to bring into being a united Usana front (Foucher 2007). The cosmetic unification of the Usana was sold to the public during the October 27, 2002, national funeral for the victims of the sinking. Although calculations for personal gains may explain some of these strategic choices, further studies are necessary to expose the specific nature of these problematic restructurings, dealings, alliances, allegiances, and coalitions.

The appropriation, substantial or nominal, of the rhetoric of tradition and indigenous religion became evident with Bertrand Diamacoune, the brother of Father Diamacoune Senghor. Bertrand Diamacoune scathingly criticized those, including the Usana, who opposed the ideological position of the MFDC in the conflict resolution (Foucher 2007), and yet he presented a younger woman, Mariama Sané, as belonging to his camp and as the president of the MFDC sacred groves.[14] Criticism erupted as to the illegitimacy of Sané as a possible leader of a sacred grove, given her youth (Foucher 2007). The lack of centralized leadership, absence of a creed, and unstructured nature of these indigenous religious practices and expressions make them vulnerable to many illegitimate uses.

Local mediators not only engage in suspect maneuverings but also accuse other women of instrumentalizing the cultural capital of the Usana. Lico's C.B. was reported as complaining in 2009, "We have organized the old women [the Usana] to intervene in the peace process. There are many who got involved for money or for doing politics . . . and they don't know anything" (Lico 2014, 135). The complaint and accusation suggest the awareness that the different actors have of one another's co-optation motives. Given the relative wealth of material, such as the studies by Foucher and Lico on the Usana of Senegal, one can access middle-level actors and their interactions with lower- and upper-level actors to understand more accurately the various deployments and abuses of social categories, specifically gender, tradition, and indigenous spirituality. Certainly, these customary female institutions in other national contexts would benefit from similar scholarly attention to forms of agentive exploitation. Moreover, although the Adjanou are the longest-standing institution in terms of the occupation of public space, few to no data exist on their inner workings and how their public and official activities are arranged and brokered.

The Return of the Native

The visibility of these institutions and their purification rituals in the western-ized postcolonial era demonstrates the persistence and even vibrancy of certain modes of being, including the exacerbated belief in the ability of ancestors, or-acles, and spirits to impact the visible modern world. That trend suggests not necessarily the otherness of the "African mind." Rather, it indexes the confu-sion about the nature of power, itself intensified by economic precarity and social disorder. The effects of economic disenfranchisement and sociopolitical embitterment also ferment moments that mobilize customary structures for re-configuring power relations. In specifically pregnant circumstances, when the modernized present is precarious and the globalized future uncertain, perhaps the safest wager is the indigenous. Some nation-states such as South Africa, Ghana, and Uganda have reasonably amended their constitutions to recognize and extend the authority of traditional leaders. The reinvocation of tradition, synonymous with the cultural turn, and the "return of the native" are also due to "an increased consciousness of indigenous cultural values as a counter to the colonial imposition of its values and the denigration of indigenous ones" (Hanna 1988, 127).

Independent of these local conditions, the return of the native also results from international interventionist practices. The recruitment of "tradition" by in-ternational aid donors within the larger investment in principles of endogamous development following the failures of the 1960–70 theories of development-cum-modernization, championed by Walt Whitman Rostow (1960) and Samuel Hun-tington (1971), is an attempt to redress the failures of the political economy in-herited from the colonial era. Specifically, in its attempts to spur democratic processes, institutions such as the World Bank championed the rhetoric of tra-dition, which calls for the incorporation of the views of "indigenous people" in development projects geared toward them. Following pressures from the bot-tom, the notion of "tradition" became appealing because of its supposed au-thenticity; its reflection of the people's views; and, finally, its affective power. The deployment of that conception in the Casamance conflict resolution in Senegal is a case in point: aid agencies, including the United States Agency for International Development (USAID), Caritas, and Germany's Weltfriedens-dienst (WFD), have encouraged the inclusion of the civil society in conflict reso-lution by conditioning their funding on gender equality and tradition. Thus, invoking tradition in the postcolonial era may be a differential repetition of the invention of tradition for colonial legitimization in the nineteenth century, as Terence Ranger (1984) argued. Such an invention, or Africanizing, in the form of defining or redefining practices results in denying them the possibility of evolving to meet new demands.

Given the female-gendered nature of genital cursing, its visibility reveals the investment in tradition, nominally or substantially, more so than in most other practices. Since the high colonial era (1880s–1910s), women have been associated with the domestic sphere (an invented tradition), the putative locus of traditional—read *African*—values, including genital cursing, which has then readily come to represent tradition. The use of defiant disrobing, considered as being of a religious nature, underscores the need to exploit certain practices. Through such practices, protesters disavow disempowering forces of modernity and oppose postcolonial necropolitics with manufactured local cosmogonies.

Since the introduction in Africa of religions such as Islam and Christianity that are associated with the history of imperialism, indigenous religions have been vilified, criminalized, and left vulnerable. Variously and inaccurately designated as superstition, paganism, and animism, traditional religions are homogeneously cast as primarily predicated on human sacrifice and devil worship, during which the initiate enters a state of frenzy, the realm of untrammeled drives and sensations. Cannibalism, extreme delirium, and convulsive dancing, a world of disorder and irrationality, supposedly mark indigenous religious practices as far removed as possible from European religious practices. Since colonial administrations could not access the inner workings of traditional institutions given the serious oath of secrecy that bound members, they exaggerated some of the practices, thus creating a fictional narrative that would support their goals of delegitimizing the practices. Delegitimizing often takes the form of witchcraft-eradication movements and attacks against so-called secret corporations that colonialists deemed sufficiently powerful to create disorder in the colony.

In response, local elites mobilized indigenous religious organizations—esoteric and initiation societies—for anticolonial struggles. Even the historiography of postcolonial nation-states engages with indigenous female and male institutions (Bayart 1989; Comaroff 1985; Ellis 1999; Geschiere 2013). The imbrication of local religious organizations in westernized African nation-state apparatuses may seem strange to those who still hold the binary distinction between religion and politics. By their very nature, these institutions belong in the public political sphere, as opposed to the formulation of religion, under its modern conception, as pertaining to the privatized and interiorized aspects of the person (Asad 1993; Foucault 1973; Ramberg 2014). For instance, in the precolonial Igbo social and political structuration schema, esoteric societies participated alongside other groupings (the council of elders and age groups) in the rule of law (Ohadike 1996). Thus, the Africanized indistinction between political participation and religious mobilization in the forms of female cursing and purification rituals was already practiced during the struggles for national liberation, and it still persists.

Several feminist and womanist-inflected works in history, political science, sociology, and social anthropology retrieved women's political involvement in the struggles for decolonization (Cleaver and Wallace 1990; Coquery-Vidrovitch 1997; Konde 2005). Whether in Cameroon, South Africa, Kenya, Mali, Nigeria, or Côte d'Ivoire, women and women's organizations took active parts in national liberation struggles, although in ways that were often illegible (feigned or real) in conventional formal political categories. These categories forcefully constructed the private sphere as the essentialized and naturalized domain of femininity and women's roles as apolitical. The public sphere was cast as unsurprisingly male-dominated, and it included treaties, wars, and power politics managed within institutions—parties, executive branches, and legislatures. The exclusion of feminized roles from high politics naturally erased women's political imprint, an erasure that feminist studies sought to politicize.[15]

In this vein, only recently did historical figures—such as Njinga of Angola, Sarraounia Mangu of Niger, Aline Sitoe Diatta of Senegal, and Nehanda of Zimbabwe—have their anticolonial contributions recognized in African historiographies, although they are often cast in mythic forms.[16] The Jola prophet Aline Sitoe Diatta of Senegal (1920–44) was arrested and sent into exile by French colonialists for her spiritual powers and abilities to mobilize people in her community. Her name has since been recruited for various and often factional political purposes. Similarly, the name and actions of Nehanda Charwe Nyakasikana (1840–98), known simply as Nehanda, one of the spiritual leaders of the Shona during the first Chimurenga (Liberation War) in Zimbabwe, recently surfaced, thus remedying the relative absence of women from contemporary politics in Zimbabwe. In a nineteenth-century town and in response to the exploitation and desecration of the land by the occupiers, the ancestors, guardians of the land, reportedly endowed Nehanda with mystical powers to lead her people into the first war of liberation. Although this war shattered the colonialists' reading of the natives as peaceful, the uprising was bloodily repressed, and Nehanda was executed at the gallows. These heroes trouble both the colonizer's historiography and the postindependence nationalist tendency to confine female resistance to the footnotes of history.

Although postcolonial bureaucrats instrumentalize indigenous female institutions within Africanized polity, these structures have proven unlikely to secure constitutional and legal recognition for the female societies. The non-formalization of their existence, despite the increasing imbrication of mystical practices in the public sphere, once more indicates the divide between state apparatuses and the aspirations and daily existences of people living in Africa and beyond.

Customary women-centered institutions and their rituals of nakedness constitute a generative site for exploring the contradictions of imported and poorly conceived forms of government within postcolonial African circumstances. These situations enable elected officials to negotiate a space of power among three exclusive systems of authority, as I demonstrate in the following scene. Using both newspaper reports and a novel, I showcase a different kind of exploitation: the use of mature female insurgent self-exposure to silence dissenting voices, as well as using the same strategy not to advance economic interests, as explored in this scene, but rather to shame postcolonial bureaucrats. The goal of scene 4 is to uncover intentional nudity as both enabling and repressive for civic leaders and for the women themselves.

In the Name of
National Interest

How was he going to impress Sikiokuu [the minister of state who is responsible for spying on the citizenry and who has enlarged his ears to serve as the Ruler's ears] *and the media without women dancers?* (emphasis added)—**Ngũgĩ wa Thiong'o,** *Wizard of the Crow*

When women strip themselves naked and stand by a major highway, that is not a peaceful demonstration. —**Kwabena Bartels**, interior minister of Ghana, 2008

In Ngũgĩ wa Thiong'o's dictator novel *Wizard of the Crow* (2006), about an African autocrat, the above rhetorical question posed by one of the protagonists—regarding the need to co-opt women's dance for both national and personal interests—exemplifies the ways in which civil leaders exploit gender and "tradition" within biopolitical postcolonial contexts. In Ngũgĩ's fictional nation-state, the dancers wearing customary garb are expected to convince the "missionaries" of the Global Bank of the people's support for the loan for building "Marching to Heaven," a latter-day Tower of Babel that the ruler believes will be the world's first "superwonder," allowing him to communicate with God. The rhetorical question suggests the long-standing nature of these exploitative practices, as I have argued in the previous scene.

Continuing the thread, this scene explores two aspects of naked agency through literary fiction and a news report. In the novel, the women's public disrobing, the shame dance, is both Africanized and secularized. Postcolonial bureaucrats Africanize the gesture to serve national as well as personal political interests. The same event also serves another purpose: the women use the recruitment of their dance to shame postcolonial officials. Another aspect of naked agency concerns how the gesture secularized to repress dissenting voices. This repressive tactic, in the name of national security, was the case of Liberian women war refugees in Ghana in 2008 (details of the incident are explored later in this scene). The dynamics at play in these events strengthen my argument about the necessity of considering the responses of the women's targets. Their reactions say more about the texture of women's agency, which ought to be regarded as open, even naked, if an accurate account of the dynamics of nakedness in protest is to be presented. Here, the complex picture of secularization shows that defiant disrobing is paradoxically both oppressive and empowering for both the women and for the authorities themselves.

Unlike news stories and their schematic structures, the novel *Wizard of the Crow* and the short story "A Trip to Rwanda" by Clarissa Pinkola Estés provide a more multifaceted map of naked agency and the ways in which disrobing goes through simultaneous and often conflicting exploitations. To do so, the narratives, like the documentary, give access to the thoughts and feelings of the women, their male targets, bystanders, and other stakeholders. In complementing and nuancing news stories, fictional(ized) accounts enable contemplation and sustained deliberation of a most intriguing gesture.

Postcolonial Agency in Legal Confusion

"Liberian Refugee Women . . . For Refugee Concerns. We are tired!!! Stop! The Humaniliation" and "Integration? No! Repatriation? Plus $1000? Yes, Yes, Yes" are written signs that preceded the 2008 alleged naked protest of the Liberian women in Ghana. Note that on one of the banners, we read "humaniliation" where "humiliation" is expected. It is unclear whether the new term is intentional. If intentional, the women were demonstrating their keen understanding of the power of words. On Monday, March 17, 2008, at 4:00 a.m., while the women and children were still asleep in the soccer field, seven busloads of soldiers arrived, accompanied by armored tanks, trucks, tear gas, and long-range weapons to arrest 630 women and children, drive them to an undisclosed location, and deport some to Liberia. The Ghanaian authorities justified the arrests and subsequent deportations: "When women strip themselves naked and stand by a major highway, that is not a peaceful demonstration" (Bartels, quoted in BBC News 2008). It is intriguing that Interior Minister Kwabena Bartels would expect a "peaceful" demonstration in the refugee camp after weeks of women's sit-ins.[1] In a government communication, Bartels threatened to forcibly repatriate all protesters to Liberia because the women's act forced the closure of schools in the camp, constituting a threat to the security of the state (R. Cook 2009). In relation to the alleged disrobing, the BBC reporter seeks to contextualize the minister's statement of "not peaceful" by highlighting the socioeconomic dimension of the women's act, writing, "Stripping naked is a traditional form of protest amongst poor and powerless women in parts of Africa" (BBC News 2008). The report mentions that several women denied exposing their tabooed body parts. That denial implies that Ghanaian authorities may have exploited the affective powers of defiant disrobing to crush the women's demands.

This instance of nominal or actual defiant disrobing was part of extended demonstrations that hundreds of Liberian refugees in Ghana staged in 2008 to demand better repatriation packages. At the time, the women protesters were living in the Buduburam refugee settlement that Ghanaian authorities founded in 1990 in response to the influxes of displaced Liberians fleeing the fourteen-year Liberian civil war (1989–2003), which had created 750,000 refugees and caused approximately 250,000 deaths. From 2004 to 2007, the United Nations High Commissioner for Refugees (UNHCR) organized unsuccessful large-scale repatriation-promotion programs (Essuman-Johnson 2011; Omata 2012). Only 7,000 refugees were resettled in Liberia, and 27,000 remained in Ghana. Many refugees considered the package too paltry to help start a life in a country where families and communities had been devastated.

Initially, male refugees contested the measures, but they were labeled troublemakers and rebels by Ghanaian authorities. On February 8, to attract more sympathy and attention, women and children, directed by the Liberian Refugee Women for Refugee Concerns, occupied the soccer field in the refugee camp to ask that they either be sent back to Liberia with $1,000 instead of the proposed $100 or be resettled in the West (United States or Europe). During the protests, the women sang songs and held signs alongside a busy highway leading into the country's capital city, Accra, some twenty-one miles away.

The dispute around whether women exposed their bodies, or whether the minister fabricated their nakedness to justify the arrests and deportations, suggests the potency of the alleged gesture as well as the uses to which postcolonial bureaucrats can put it. If there were no case of shaming self-exposure, Ghanaian authorities may have manufactured it, using the language of national security, in order to abort the protest demonstration. Echoing other bureaucrats (as I show later) is the suggestion to protesters of other and more effective ways of prosecuting grievances. As I have now established, most instances of defiant disrobing are subject to multiple interpretations. Therefore, Bartels's exploitations (nominal or substantive) of the supposed aggressive nakedness enable him to negotiate a space of power among three exclusive systems of authority:

the language of indigenous practices (customary/cosmological); the economic and political instability of the postcolonial nation-state; and the pressures of former colonizing powers, masquerading as the international community with their legal arrangements regarding refugees (biopolitical–postcolonial). These systems need to be understood both on their own terms and in competition with one another.

In using his customary framework to recognize the implications of the women's supposed gesture as dangerous, the minister's response proves the legibility of the women's public disrobing outside their country of origin, Liberia. This recognition belies the artificiality of the nation-state boundaries between Liberia and Ghana bequeathed by the *mission civilisatrice* (civilizing mission) of colonialists. Bartels's awareness, the result of shared cultural practices and their legibility across national borders, supersedes the constructedness of state boundaries at the same time that it reveals the remnants of shared precolonial and colonial episteme. Had the minister remained within the customary framework that he used to interpret the act as not peaceful, he would have been obligated to meet the protesters' demands as prescribed in the script provided by authorizing cosmologies available in ethnographies. Further, his cosmological interpretation-cum-reaction demonstrates that the minister is not immune to the language of mature women's public undressing. To consider the arrest and deportation as the only official male reaction is to disregard the ways in which the deep-seated influence of cultural traditions undergirds most practices in so-called modern cities. Colonialism and neocolonialism cannot have completely erased centuries-old beliefs. It is possible that the violent public nationalist interpretation can be different from the private reaction, wherein the public official may acknowledge the possibility of being shamed and/or cursed and would consequently seek protection.

Being unable to fulfill the women's demands, the minister quickly stepped out of the customary framework in order to adopt legal language—that of arrest and deportation—pertaining to the state, without fully complying with relevant international protocols. The possibility of switching frameworks, resulting from the minister's difficult position of recognizing multiple sets of practices, shows the paradoxes for women of using this form of contestation outside kinship networks and in the postcolonial city.

The second interpretation/reaction to the Liberian women's incident is one that I call "biopolitical–postcolonial," which draws on multilayered explanatory accounts that concern the legal category of indecent exposure, the women's status as refugees, and the liberty that the minister took to arrest and repatriate them. The first Ghanaian president, Kwame Nkrumah, and his ban on nakedness among supposedly naked or scantily clad northerners in postindependence Ghana, a point I elaborate on later, inform the biopolitical–postcolonial interpretation. The shadow of this policy has hovered over Ghana's postcolonial social and political history of clothing and perhaps informed the category of "indecent exposure" in the Ghanaian legal code. A clause in Article 25 of Ghana's Criminal Offences Act criminalizes public indecency, which consists in "willfully and indecently exposing oneself in any public place or in view thereof, or exposing one person in any place with intent to insult any person" (Cammaert 2016, 185). As in most Anglophone countries in Africa, this clause is a residue of the colonial era "repugnancy clause" (Hoad 2007). At the time, the clause prohibited customs and practices—specifically polygamy, witchcraft, and rituals—that the colonial administration deemed immoral, backward, and insulting to good conscience. In general, two legal systems, customary law and metropolitan law, were operating simultaneously to regulate social issues during the colonization of West Africa. Unlike the metropolitan law, the customary law that settled legal matters among the natives was still regulated by the colonial authority and its modes of judgment. Even two centuries after the end of direct colonization, postcolonial nation-states still carry the residual value judgments of the former colonial empire.

The Liberian women protested in the paradigmatic biopolitical space—the African refugee camp—with the suspension, heightening, and confusion of most legal and political prerogatives (Agamben 1998; Arendt [1948] 1973; Hussain 2007). For instance, in her press briefing about the arrest, UNHCR spokesperson Jennifer Pagonis chastised the Liberian refugees for their refusal to use existing channels in Ghana to address their grievances. She reminded the protesters of their "obligation to obey the laws and regulations of their country of asylum" (UN News 2008). Yet it remains unclear for how long these refugees must use "normal" channels of claims making in order to secure more than a dismissive reaction from their targets. Further, what the women consider in these circumstances as called for and thus normal is their insurgent nakedness. Thus, what is normal is already a contentious point. More confusing in the handling of the prosecution of these grievances is the minister's legal language. He called the protest demonstrations "a breach to public order" as stipulated in the 1994 Public Order Act (Ghana Legal 1994). Specifically, he says, "These acts are clearly against the laws of Ghana and the government wishes to advise all refugees, those involved in the illegal demonstration at the Buduburam camp that it has a duty to maintain law and order in the country" (Ghana Web 2008). Despite the invocation of the Public Order Act, the statement contains many legal inconsistences because the police, and not the military with heavy artillery, have the prerogatives of imposing a curfew, dispersing protesters, and even arresting suspects. The order does not endow the police or the Ministry of the Interior with the power to deport without legal due process. In other words, in the case of illegal activities, the legal system should take control of the case and duly prosecute the accused; it is not the prerogative of the minister of the interior to take matters into his own hands and deport the refugees or threaten to do so.[2]

The Ghanaian authorities' response is what I call "mild secularization" because it acknowledges the significance of disrobing without complying with its implications. This mild secularization is paradoxically oppressive and emancipatory for both the authorities and the women. By stripping themselves of clothing, Liberian women are framing themselves as victims of both Liberia and Ghana, while expressing their rights and revealing the lack of state protection. The refugees become figures akin to Agamben's "naked lives" to whom governments are not extending political rights (1998). Perhaps, after living for years in a refugee camp with poor sanitation, insecurity, and precarity as their exceptional but now normalized conditions, these women lost faith in the conventional channels of political contestation. Their resistant self-exhibition provoked a somewhat counterproductive effect: the spokesperson of the UNHCR described the protesters as "increasingly threatening and disruptive" (Ghana News Agency 2008), and the official response of the Ghanaian government was to arrest them. To unveil tabooed body parts either to reveal their precariousness or to inflict injury onto the body politic is one of the forms that contestation can assume in this context. The women challenge us to rethink our assumptions about their bodies' inabilities to act politically. By resorting to their bodies, which are all they have left at the time, these women refuse to comply with the demands of modernity and its normative modes of prosecuting grievances. Thus, naked agency becomes preemptive and a way to take back agency in a heightened scene of vulnerability while also denying temporarily the women's targets any agentive power.

The Shame Dance and Immunity to Shame

In the novel *Wizard of the Crow*, inflicting pain on the body politic is the plan of the women whom the minister recruits to perform their "traditional" dance as a sign of the people's support of the dictator's outrageous, costly, and self-aggrandizing project. Ngũgĩ wa Thiong'o's novel is a valuable and internationally circulating postcolonial literary narrative of African womanhood that includes defiant self-exposure.[3] Written originally in Kikuyu, *Wizard of the Crow* was translated by the author into English and published in 2006 to rave reviews for its orality-infused storytelling, its largely fantastical bent, and its critique of power in the aftermath of white rule (Forna 2006; Jaggi 2006; Turrentine 2006; Updike 2006). The 766-page novel is a story of both oppression and resistance to oppression, staging many affective dimensions—shame, anger, joy, disappointment, and fear.

The sprawling tale of tricksters and magic, heroic escapes, and outrageous somatic diseases exposes the plots of the Ruler's deluded and sycophantic ministers to keep the despotic rule intact, as well as those of their opponents, members of the underground resistance movement of which Kamiti, also known as "the wizard," is a part.[4] At the heart of the novel is the Ruler petitioning the Global Bank for funds to build the world's tallest building—the absurd and doomed "Marching to Heaven." Once a petty bureaucrat, the mighty Ruler comes to power in Aburīria with the aid of the United States, which helps depose the British-backed dictator. But the machinations of the Ruler fail to secure the loan from the Western lenders because the resistance models of the people increasingly challenge his oppressive mechanisms. Frustrated, the Ruler goes on a diplomatic mission to New York City with the goal of convincing the lenders, who recommend that he return home to attend to the increasingly bold resistance movement of Nyawira and the people. Consistently successful at embarrassing the despotic ruler, Grace Nyawira is Kamiti's girlfriend and the leader of the democratic Underground Movement for the Voice of the People. After one grotesque story, rumor, and machination after another, the Ruler is finally deposed by Tajirika, the new minister of defense, who plots with Wonderful Tumbo, the head of the Santamaria police station. Not surprisingly, the overthrow of the Ruler fails to bring Aburīria the long-awaited freedom from autocracy, mismanagement, and corruption. Tajirika declares baby D (democracy) dead, and the country an empire. He even names himself Emperor Titus Flavius White Head. True to his ideological position of opposing dictatorship, Ngũgĩ ends the multigenerational novel with Nyawira and Kamiti promising to continue fighting for freedom.

The novel's underground movement mobilizes several innovative strategies using the female gender, the precariat, and animality: the dissidents assume beggarly appearances during official visits by American investors; the women create a parallel justice order wherein they kidnap, judge, and sentence wife-batterers; and they cause panic during political gatherings with plastic snakes. The most impactful form of resistance that the insurgents marshal is shaming the Ruler with mature female self-exposure. Similar to the exploitation of women dancers in customary garb for factional political mileage, as with the Adjanou in Côte d'Ivoire, Kaniuru, the postcolonial bureaucrat in the fictional Aburīria, recruits women dancers to grace the ceremony to make the headlines of newspapers and news reports on television stations around the country. Headline making carries significant political influence against opponents and with international donor agencies interested in gender and tradition, as the organizer's rationale: "*How was he going to impress Sikiokuu* [the minister of state who is responsible for spying on the citizenry and who has enlarged his ears to serve as the Ruler's ears] and *the media without women dancers?*" (Ngũgĩ 2006, 306; emphasis added)

However, the dancers come with a plan: to perform the shame dance to punish the Ruler for incurring more debt to finance the unwanted project. They do so by creating a scandal that ruins the ceremony. The subversive intervention of the women crystallizes the reverse exploitation of gender and tradition. In the middle of their procession and dance, punctuated with undulating hip and hand movements, the women stop according to their plan and suddenly face their audience, with their backs turned to the platform.

> All together we lifted our skirts and exposed our butts to those on the platform, and squatted as if about to shit en masse in the arena. Those of us in the crowd started swearing: MARCHING TO HEAVEN IS A PILE OF SHIT! MARCHING TO HEAVEN IS A MOUNTAIN OF SHIT! And the crowd took this up. There were two or three women who forgot that this was only a simulation of what our female ancestors used to do as a last resort when they had reached a point where they could no longer take shit from a despot; they urinated and farted loudly. Maybe need or fear overcame them, or both. (Ngũgĩ 2006, 250)

The disrobing gesture causes a commotion in the crowd, whose uncontrollable movement pushes the platform that holds the Ruler and his guests. In reaction to the women's act, the police await the order to shoot the women, an order that the Ruler does not grant because of foreign journalists in attendance. What happened in Igboland during the colonial period (1929) would have happened in Aburĩria if not for the presence of foreign news reporters and their cameras. Here, the women employ colonialist tools (cameras) to their benefit despite the conventional power of the tools to oppress (photographs of African women circulating in Europe during colonialism). After the dance, the details of which differ from one account to another, the platform that holds the Ruler and his guests begins to sink into what some accounts consider to be a muddy pool made of a mixture of feces and urine.

The feeling of being shamed, of being defied, constitutes a trigger for more oppressive violence, although the women's shame dance does send the Ruler into hiding for seven days, seven hours, seven minutes, and seven seconds (253).

The punitive gesture is defined in multiple ways: "shameful acts" (242, 244), "shameful deeds" (336, 505), "shameful betrayal" (378), "acts of shame" (336), "a shameful scandal" (218), and "the drama of shame" (744). Intolerable to the dictator is that the shaming occurred in front of what the narrator calls the "missionaries" of the Global Bank, members of diplomatic missions, and cabinet members in a historic gathering that the country had not witnessed until then. Along with shaming the Ruler (135, 136, 235), the women of the fictional country's Underground Movement are also said to have "brought shame" on the Ruler and the nation, with which he has become one (219, 225, 301, 347), as well as on "the site of Marching to Heaven" (236).

The importance of shame as a structuring principle explains the astronomical number of times the term *shame* appears in African fiction and in proverbs.[5] One example is the Malinké proverb "Saya ka fisa maloya ye," which is often translated as "Death is preferable to shame" (Bailleul 2005, 89). Likewise, Ebiere, a prominent character in John Pepper Clark-Bekederemo's *Song of a Goat* (1961), reiterates similar fears of shame, moaning, "Oh, how I wish I'd die, to end all this shame, all this showing of neighbours my Fatness when my flesh is famished" (2).

The potency of shame in the worlds of the play and the proverb can best be understood within the context of a society that has constructed its values around communitarianism, a social organizing principle predicated upon the belief that the formation of the self is dependent upon the cues of others. Thus, shame—the severing or threatening of social bonds, the fear of having one's social status threatened with the resultant ostracism and rejection—is worse than literal death. It consequently constitutes a tremendous motivating force that causes one to abide by prevailing shared moral codes and socially sanctioned behaviors. Such codes require that the member be bound to the community, so much so that the construction and deconstruction of her identity is overwhelmingly accomplished in tandem with the moral approval and disapproval of other community members. That the preference of choosing death over shame is expressed as a proverb speaks to its axiomatic qualities.

Ironically, the Ruler considers others shameless when, in his phantasmagorical and delusional aspirations, he is the shameless one within the context of the more encompassing sense of shame. Shame in this sense consists of the lack of humility, discretion, discernment, respect for elders, harm avoidance, and adequate public comportment and posture.[6] It is in fact the Ruler's shamelessness that translates into his corrupt, degenerate, and oppressive rule, for which the female members of the resistant movement decide to teach him a lesson by shaming him. The after-effects of westernization—cultural diversification, detraditionalization, urbanization—afford a certain kind of agency to postcolonial rulers who are only mildly inconvenienced by shame.

A correlation exists between the work of the female shaming institution, the Ruler's reactions to being shamed, and narrative form (see Diabate 2019). Specifically, the narrative includes the repeated terms *shame* (17), *shameful* (9), *shamed* (2), and *shameless* (3). In his exploration of shame and its influence on social reform, Ngũgĩ positions women and their socially sanctioned shaming institution to wield this kind of power. Though consistent with historical accounts of women's forms of conflict management, this view is unconventional in our globalized understanding of femininity as the repository of shame. Menstruation putatively throws the girl into the realm of potential shame as "she bears within her body the seeds of sexual shame in such a way that this feeling about herself detaches from any act she may or may not commit, for shameful acts can merely confirm this dreadful self-knowledge" (Johnson and Moran 2013, 2).[7] From the published literature, however, women are either framed as incapable of feeling shame or endowed with the capacity to produce shame in others. It is unclear if and how these accounts explain why recently in South Africa, Malawi, Kenya, and Zimbabwe, crowds have publicly stripped and humiliated women in the streets and in online spaces for their supposedly shameless sartorial and sexual choices (Ligaga 2014, 2016). These shaming assaults are undoubtedly disciplinary techniques deployed to inscribe heteropatriarchal ideals that attempt to normalize female corporeality and sexual practices.[8]

In the novel, the organizers of the dance are elated because the Ruler, who has at first proven exceptionally untouchable, was "driven away by Women Power" (253). However, the effect of the shame dance, while powerful, is also only temporary because the Ruler and his cabinet succeed in constructing an official account that curtails the power of women's agency. Similar to most events of aggressive flashing, the incident from the novel described earlier elicits multiple and often conflicting reactions from at least four categories of actors: (1) the Ruler and his sycophants, (2) foreign dignitaries, (3) the women dancers and other dissidents, and (4) the people. The foreign bankers, for instance, laugh, mistaking the flashing of buttocks as a "humorous native dance." However, in light of the Ruler's serious reaction, they then understand it to be a "solemn native dance" (250), an expression of "black humor from an ancient Aburīrian ritual" (252). Most people in the crowd are confused by the gesture. The official version by the Ruler and his cabinet explains the women's dance as "a sacred Aburīrian dance performed only before most honored guests" (242). In relation to the generalized African context, terms such as *ritual*, *sacred*, *native*, and *ancient* conjure up ideas of both authenticity and backwardness. Ngũgĩ plays on all those registers by reimagining these multiple and often contradictory exploitative moves.

Because of the zealotry of the Ruler, the practice commanded only a mildly effective response: driving him into hiding only temporarily. The Ruler's reactions are consistent with conventional descriptions of shame reactions in dominant Western contexts. However, according to nondated oral accounts in anthropological studies, genital shaming of this type would have unseated the dictator (Ardener 1973; Conrad 1999; Prince 1961; Ritzenthaler 1960). The temporary and mild inconvenience suggests the dwindling power effects of mature female public shaming.

Although the Ruler does not respond in conventional ways to shaming, he is not immune to the shame dance, which suggests sympathy for his personhood despite his authoritarian and repressive political practices. Indeed, those immune to female shaming will not react according to certain expectations (conventional shame reactions) in the face of women's punitive act. These members are considered dangerous elements to communal commerce, as already argued.

That the gesture confuses most people in the crowd speaks to an important aspect of naked agency: the precariousness of defiant disrobing in the postcolonial city. The limits of body language, even that of exposed tabooed body parts, necessitate the intervention of spoken words, which are also apparently insufficient; in *Wizard of the Crow*, Nyawira reports that several people in the crowd start swearing, "MARCHING TO HEAVEN IS A PILE OF SHIT! MARCHING TO HEAVEN IS A MOUNTAIN OF SHIT!" a chant that the crowd takes up (250). Perhaps the confusion in the crowd stems from the ethnic diversification in postcolonial cities and the existence of multiple customary practices.

This retributive self-exposure is not of an indigenous religious kind. As the narrator clarifies, the women's act was "only a simulation of what our female ancestors used to do as a last resort when they had reached a point where they could no longer take shit from a despot" (250). Why is it only a simulation? Is it because biopolitical circumstances in Aburĩria, where the distinction between the dead and the living is tenuous, do not warrant such a response from the women?[9] This specification puts pressure on the arguments that consider most dramas of nakedness by mature women in protest as a form of genital cursing or shaming in the religious sense. Certainly, there is shaming here, but it is to expose the Ruler's mismanagement of resources, and not to directly unseat him.

Further, contrary to women's deepest expectations, their shaming gesture causes the Ruler only a temporary inconvenience. Even more counterproductive to the women's gesture is the fact that the autocrat, in the midst of hiding, organizes his vengeance.

> Vengeance is mine, sayeth the Lord, and was he not the Lord of all Aburīrian women? Yet no matter how hard he considered the matter, he remained unsure as to what to do or where or with whom to start the vengeance like that which he had shown Rachael [his legal wife]. Unable to act, his torturous thoughts always returned to the treacherous drama at Eldares in which the *women shamed the nation* before the eyes of foreign dignitaries and, worse still, in front of the Global Bank missionaries. For days on end, after the drama at Eldares, the Ruler kept to himself, trapped in secrecy. (135, emphasis mine)

One of the most notable reactions of the Ruler is to construct the women in an ambiguous fashion in that they are both shameful and shameless: "dancers of shame" (251), "shameful dancers" (423), "shameless women" (263), and "shameless primitive dancers" (376). The Ruler's insulting rhetoric of primitiveness indicates the paradoxical nature of women's strategies. That paradox, which has been consistent in responding to mature female naked protest, raises one of the framing questions of this book: How effective is the use of symbolic resistance strategies that are potentially misunderstood? The link between primitiveness and shamelessness that the Ruler establishes is unsurprising given the recurrent trope in colonialist racist judgments of the natives (C. Achebe 1970; Fanon [1952] 2008; Kemedjio 1999). Thus, for some, the women's act reflects their inability to adopt civil/political ways of prosecuting grievances (F. Ajayi 2010).

However, most studies of punitive and purifying rituals overlook possible backlash. Perhaps the need to construct women as endlessly powerful in order to uphold an indigenous mode of being, thereby complicating commonplace images of women as victimized by their traditions, accounts for the limited attention to the backlash against collective female defiant disrobing.

Access to the interiority of the women's targets and of other stakeholders makes the fictional genre one of the most important venues for a richer account of naked agency. *Wizard of the Crow* and T. Obinkaram Echewa's *I Saw the Sky Catch Fire* mobilize similar narrative strategies and multiple narrators for such access.[10] In the former, for instance, the recounting of the shame dance and of its effects requires several narrators. The main narrator establishes the atmosphere preceding the shaming gesture as one of awe, silence, and deference. The women's procession is described as solemn, dignified, awe-inspiring, and respectful, among other adjectives (250). However, a dramatic tone replaces the solemn one, as narrated by the lead organizer of the resistance movement, Nyawira.

Given her keen interest in the event, her account provides a detailed description that ensures that the choreographed nature of shaming self-exposure is not lost on the reader.

In addition to providing access to the targets' interiority, the deployment of multiple narrators also functions as a form of resistance to the dominance of elite sources and voices. Drawing on Luise White's theorization of rumor in *Speaking with Vampires: Rumor and History in Colonial Africa* (2000), I follow Robert Colson's (2011) insight to consider multiple narrators as a form of rumor—of nonelite sources narrativizing stories and thus creating realities in the diseased postcolonial nation-state. This narrative technique seems necessary in order to highlight the powerful nature of this form of resistance or punishment.

Through the dictator novel, Ngũgĩ provides a more expansive understanding of shame in the postcolony and how it reflects the ever-changing nature of oppressive power and resistance. The ways in which shame is meted out by opponents and how it temporarily inconveniences (but ultimately fails) to unseat the brutal dictator suggests Ngũgĩ's abandonment of the complex hero characters of his previous novels. Ultimately, the narrative demonstrates the shifting grammar of shame in Ngũgĩ's writings and, relatedly, his espousal of a more moderate view of how positive sociopolitical change should unfold.

Clothing Nakedness

Ngũgĩ's Ruler feels shame not because of being exposed to his fellow citizens but because of the presence of foreign dignitaries. This accountability to a foreign order is not unique. The shame that leaders and intellectuals express about female disrobing often speaks to the refusal of signs of so-called backward and fetishistic practices to die. Such reactions to naked protest have occurred in The Gambia, Nigeria, and South Africa. Nowhere was this hope to uphold a specific national image more evident than in the late 1950s' and early 1960s' antinudity campaign in Ghana. Nakedness in the northern regions of Kwame Nkrumah's newly independent Ghana was not for purposes of contestation but rather a mode of being among the peoples. Concerned about the international reputation of the country, then considered within the African diaspora as the Black Star, the staunch social and political activist, philanthropist, and women's leader Hannah Kudjoe—often at odds but also in tandem with Nkrumah's government—sought to clothe northerners. The government and NGOs strategized campaigns to clothe the culprits; eventually they threatened to arrest all naked or scantily clad northerners. The stated goal of this project was to liberate northerners from the clutches of "backwardness" and "savagery" (Allman 2004; Donkoh 2015; Sarpong, Sarpong, and Botchway 2014). Although Ghana's antinudity campaign took place in the immediate postindependence years, it still resonates with the goals of current African intellectuals and political leaders who seek to uphold the image of their countries as modern and thus freed from "darkness." In this context, naked agency erupts as the return of the repressed to smear those self-serving and self-promoting campaigns.

In 1977 Tanzania saw a similar project of clothing supposed nakedness. The unpopular government decree sought to clothe adult Maasai according to European sartorial codes if they were to use public transportation. However, the measure led the Maasai to strip naked in protest. According to Dorothy Hodgson (2017), American sociologist Colby Hatfield witnessed the collective naked self-exposure and described the women as upset because of the inaction of men against the ban. The women in the south genitally cursed their local Maasai leaders; those in the north cursed the national parliamentarian and collected money to prepare a trip to Dar-es-Salaam, the capital city. Eventually, a ritual slaughtering of several cattle to propitiate the spirits lifted the curse, which persuaded the women not to march on Dar (Hodgson 2017). The Maasai's stripping naked to protest against a ban on nakedness speaks to the differing understandings of clotheslessness.

Civic leaders' conflation of the deliberate absence of dress with backwardness indicates adherence to modernized and dominant Western sartorial norms and codes of behavior. Nakedness is in the eye of the beholder (Berger 1972), and what counts as clothing, that "social skin" as Terence Turner (2012) calls it, is subjected to shifting and often contradictory significations. The degrees and kinds of exposure, what Erving Goffman calls "the *orientational* implications of exposure" ([1963] 2008, 211), make it almost impossible to speak of nakedness without defining its contours. Unlike in the dominant Euro-American context, shaped predominantly by Judeo-Christian precepts, lack of dress in African contexts of genital cursing and shaming is not incompatible with modesty. Lack of covering is not suggestive of obscenity or indecency, which postcolonial bureaucrats claim.[11] From the perspective of northern Ghanaians and the Maasai, another civilizational account of dress was operative. For these natives, loincloths made with bark, and cloaks for adult males and females in public settings, constituted full clothing.

Further, the existing literature suggests that in several communities, the state of undress is valenced both positively and negatively. Nakedness is positively valenced as youth, innocence, purity, beauty, and health (Bastian 2005). For instance, among the Warego of the Belgian Congo, a new father of twins strips naked and stands in front of his hut with his genitals bared as a sign of fertility and health (Ombolo 1990). The residual effects of these native under-standings of what counts as clothing exist in certain postcolonial societies, in-cluding among the Nnobi of Nigeria. When Ifi Amadiume attempted to pho-tograph the fully clothed female chief, described as a "male daughter" in *Male Daughters, Female Husbands* (1987), the chief respectfully declined because she was not "dressed." In her article "African Women's Body Images in Postcolonial Discourse and Resistance to Neo-Crusaders" (2008), Amadiume writes, "She told me that she was not dressed; that she was barebodied! I found this puz-zling since I could see that she was fully dressed wearing a blouse and wrapper" (55). During another visit, the chief came out topless for the photo shoot. The anthropologist stated, "I needed to quickly make sense of our differing under-standings of the meaning of nakedness, plainness, and being dressed" (55). The female chief's understanding of being dressed according to her sociopolitical rank and privilege differed from Amadiume's.

In indigenous religious settings, among actions such as oath taking, initia-tion, healing, purification, and protection, the deliberate removal of clothing signifies positive characteristics, including honesty, absolute loyalty, and pu-rity of the heart and of intentions.[12] Paul Hazoumé's (1956) study of the Fon of Dahomey (the Republic of Benin) observes, "We believe that nakedness is required [in oath taking] because to wear clothing indicates that one is tied to something. . . . The oath takers also want to signify by their nudity the right-ness of the heart and the good faith which reigns in rapport with their commu-nity" (100).[13] Drawing from Hazoumé's study, Suzanne Blier (1995) adds that nakedness among the Fon is also "identified with each person's spirit persona (sε), the incorporeal essence responsible for one's life, morality, and destiny [core self]. [Thus], the word for nudity is sε gbeji, literally 'the voice of [gbeji] of sε,'" which also stands for virginity (169). Given that nakedness can signify purity, clear conscience, and originariness, it also suggests spiritual transition because in the voyage from the material to the nonmaterial world, one passes through the state of undress. One cannot access the world of the ancestors when still saddled with worldly attachments and problematic emotional trappings (Blier 1995; Hazoumé 1956).

The nexus between nakedness and spiritual transition, however, becomes complex with individuals suffering from a mental illness, which explains why an insane person frequently goes naked in public in past Fon and Igbo socie-ties. Because clothing is associated with sociality—external markers of social distinction, belongingness, and reciprocal relations—stripping naked amounts to literally as well as metaphorically severing relations by discarding one's social identity as husband, wife, sister, brother, daughter, son, uncle, aunt, mother, or father (Bastian 2005).[14] The signification of nakedness as lack also finds relevance in socioeconomic terms as it is often coded as weakness and impoverishment. Among the Fon, artists represent those in position of wealth and privilege through clothed sculptures or figures adorned with dress-associated signifiers, and they use nakedness and near-nakedness to suggest a situation of relative impoverishment (Blier 1995). The contradictory nature of nakedness makes it a powerful mode of contestation. The inconsistencies also suggest that Africa is not a country, it is a continent with diverse cultural practices and colonial histories.

Annoyance with and need of postcolonial leaders to dress supposed nakedness resonates with unfounded accusations that naked agency is exposing images of African women's bodies to exploitation. A few African academics in the United States have said that to analyze mature women's defiant self-exposure is to subject these bodies to double injury: racist colonial texts and texts in the current era by neocolonialists in search of exotic images, as well as texts by the rescue industry, have physically and epistemically violated African women's bodies for centuries. In short, these self-appointed gatekeepers saw unquestionable similarities between the ubiquitous discourses on female genital surgeries and my exploration of genital cursing and naked protest.

For instance, in 2012, during a public lecture at a major American university, a professor of African philosophy and international relations challenged me to explain—convincingly, he specified—why we needed to expose the genitalia of our mothers yet again to the voyeurism of Westerners. My short answer to his objection was that highlighting these sources of insurgent disrobing was a way for me to make sense of the ways in which literary fiction and critical work portray my body through the ubiquitous images of violation, rape, and cutting. Further, the proliferation of events involving female nakedness and its dissemination through social media necessitated examining naked agency for what it says about Africa in the biopolitical era.

Before these postcolonial goals of clothing naked bodies literally and symbolically, the desire to do so was, of course, already present in the colonial era. Examples of the latter are portrayed in the American psychologist Clarissa Pinkola Estés's *Women Who Run with the Wolves: Myths and Stories of the Wild Woman Archetype* (1992), a collection of reinterpreted folklore that presents mythical stories and fairy tales of women's power in different cultural and historical settings. "A Trip to Rwanda" is a three-page vignette that describes how Rwandan women defiantly lifted their skirts in protest against their colonial governor, who forcibly clothed them in the fabric of civilization and made them stand along a dirt road to welcome General Dwight D. Eisenhower during his alleged World War II visit to American troops stationed in Rwanda.[15] Differing readings of sartorial codes lie at the core of the incident.

In the vignette, the native women's daily dress code was no more than a necklace of beads, and sometimes a little thong belt. Therefore, in preparation for the visit of the "white" man and to protect the sensibility of the visitor, the governor consulted with the village chief to provide women with skirts and blouses. Contrary to what the governor thought, these Rwandan women's bodies were never naked, because their hairdos, body and facial markings, and other bodily signs were texts covering them. By literally and figuratively stripping women of their beads and necklaces and dressing them in skirts and blouses, the governor attempted to treat them as mere bodies in the service of the colonial regime. But, as Eisenhower's jeep approached, the women lifted their skirts to cover their faces, seeking to shame their tribesmen, the governor, and his honored guest. Lifting their skirts became a moment of resistant agency for the women. With this gesture, they signaled that they could be clothed and forced to stand by the roadside, but they could not be disciplined as docile objects. However, the bodily performance of agency necessitates that the subject and the regime of control share in the same episteme.[16] Here, that sharing was disrupted because the governor and Eisenhower regarded the women's gesture as an expression of sheer backwardness—the women's supposed unfamiliarity with full-body clothing.

Whether the colonialists understood or misunderstood the incident, it succeeded in inscribing women's capacity to react temporarily as they disrupted the decorum and protocols of the parade. More broadly, these moments of resistance are empowering. However, they depend on two crucial factors: the identity of the observers and the nature of their responses.

One such response is that of co-optation, with I explore in the next scene through the postmodern film *Les Saignantes* (Bekolo 2005). In my close reading, I situate *Les Saignantes* within an emerging visual archive of insurgent nakedness in Africa and demonstrate how the film exploits women's purifying ritual to heal the postcolonial modern state. Through women's bodies and the film's tropological counterdiscourse to postcolonial necropolitics, in the deathscapes hope becomes viable against the further implementation of dystopia.

Film as Instrumental
and Interpretive Lens

Filming in Africa means for many of us
Colorful images, naked breast women, exotic dances, and fearful rites.
The unusual.
Nudity does not reveal
The hidden
It is its absence
—**Trinh T. Min-ha**, *Reassemblage*

As Trinh T. Min-ha noted in 1982, the relationship between Africa and filming is one of strife. The cinematic genre's obsession with "naked breast women,
exotic dances, and fearful rites" from Africa answers the need to reveal, to understand, and to control that which is supposedly unfamiliar, hidden, and
threatening. If that is the relationship between film, visual art, and a supposedly naked Africa, how should we read the dynamics between cinema and purifying rituals in the postcolonial context? How do other visual productions on
protest nudity participate in this problematic debate? These are the overarching
questions that this scene seeks to answer. I argue that Jean-Pierre Bekolo's feature film *Les Saignantes* (2005) mobilizes women's images of purifying rituals to
heal the diseased postcolonial nation-state—an appropriation of the women's
society that is not the direct political co-optation of their organizations by politicians and other power brokers, as I have laid out in the previous scenes. Like
three other films on assaultive nakedness—*Uku Hamba 'Ze (To Walk Naked)*
by Jacqueline Maingard, Sheila Meintjes, and Heather Thompson (1995), which

I discussed earlier; Abigail Disney's *Pray the Devil Back to Hell* (2008); and Candace Schermerhorn's *The Naked Option: A Last Resort* (2010)—Bekolo's film carefully desexualizes the purification ritual to avoid inflicting double injury on African women via images.[1] However, desexualization weakens the potency of genital purification via genital cursing.

One of the best-known West African films, *Les Saignantes* is Bekolo's scathing political critique of the postcolonial ruling class. The postmodern film features two young women who exchange sexual favors with highly influential male state officials for business contracts. Unbeknownst to the young women, the spirit of the Mevoungou takes possession of their bodies to execute the social and political mission. The film does more than comment on the failed dreams of decolonization; it also reimagines images of women's institutions, turning them into purifying, empowering entities for communal healing. The use in the film, however, verges on instrumentalizing women's institutions in a space where women have lost many of their prerogatives because of the restructuring of gender norms.

The Mevoungou are a powerful precolonial female secret society of the Beti in Cameroon. The members' power is predicated upon the clitoris, which the women feed, massage, and worship during secret rituals of purification and cursing. The cinematic adaptation and appropriation of the Mevoungou are compelling because nineteenth-century German missionaries abolished most Beti indigenous female secret societies and their rituals, including those of the Mevoungou. The Beti revered the society. When the group's rites were abolished, Beti women were deprived of some of their cultural avenues of communal prosperity and empowerment. These palimpsestic layers of discursive and nondiscursive practices have pathologized and attempted to silence the women.

In this scene, I first position the film within Bekolo's filmmaking and explain how the idea for the movie came from an actual exasperating life event. Drawing on ethnographies and anthropological sources, I provide an overview of the female society. An analysis of the nighttime setting in the film is useful because it underscores not only the necropolitical and apocalyptic vision of an annihilated state but also the necessity of enlisting a female-centered ritual to heal the ills of the found(l)ing fathers. The circumstances that demand the intervention of the collective cleansing ritual by the society include drought, pandemics, calamity, and war. Here, I show how the filmic genre imagines a space where human political agency is almost nonexistent as men and women, through their behaviors, become animal-like. These figures in the film are reduced to naked lives—denied political prerogatives and exposed to death with impunity. *Les Saignantes*, and its ability to express viscerally the manufacturing of naked lives, is powerful in communicating the kinds of circumstances that beget genital cursing and purification within postcolonial necropolitics.

This scene examines the purifying capacities of women's societies to highlight decapitation, castration, cannibalism, and theophagy as the film's tropological counterdiscourse to the further implementation of dystopia. But this kind of imagining hope also creates paradoxes as it continues to use images of women's bodies while ignoring their concerns. This dynamic demands a different conceptualization of women's action.

Les Saignantes and Bekolo's Cinema

Les Saignantes (English titles: *The Bleeders*, *The Bloodettes*, *The Bloodiest*) was produced in 2005. Previewed at the 2005 Toronto International Film Festival (TIFF), the film received unfavorable reviews. In 2007 it was previewed at the sixtieth Cannes Film Festival and the twentieth Festival pan-Africain du cinema et de la television de Ouagadougou (FESPACO, Pan-African Film and Television Festival of Ouagadougou), where it won the Etalon d'argent de Yennega, the second prize.[2] Despite its international exposure, the film failed to find a distributor until May 2009, when it was distributed in France.

Set circa 2025 in an unnamed postcolonial African country, this postmodern horror, science fiction, and crime film opens with a sexual encounter between a sexy woman, Majolie (Adele Ado), and a high-ranking official, known as the SGCC (Secrétaire General du Cabinet Civil). The encounter results in his death. Panicked and frustrated over the loss of her sexual investment, Majolie (unaware that the Mevoungou possess her body) calls her best friend, Chouchou (Dorelia Calmel), to help her cover up the terrifying incident. After their abuse of the corpse, les saignantes take the body to the butcher shop and attempt to convince the animal-like butcher that it is a "load of fresh meat, prime beef." Miraculously, the butcher recognizes the SGCC's body by tasting it, then proceeds to separate the head and testicles from the trunk, handing the former to the protagonists. Confident that they have hidden the incident, Majolie and Chouchou then realize that they may land other business deals during the SGCC's funeral. They promptly devise a plan to reconstitute the body so that the SGCC's funeral may be organized. In the middle of the night, they visit the mortuary and bribe an alcoholic middle-aged mortician, who provides them with a body that matches the head.

With a reconstituted body, the SGCC's family and state officials organize his funeral, but before long, the SGCC's wife realizes that the body is not that of her husband. As for the protagonists, in order to strike a business deal, they attempt to seduce the minister of state (Emile Abossolo Mbo). An intelligent man who is both a sex maniac and knowledgeable about the occult, the minister proves to be more insightful than they anticipate, and he challenges them. Throughout the film, the young women go through several purification rituals that strengthen them. Finally, with the help of the Mevoungou, they use their supernatural powers in a type of martial arts dance to defeat the minister of state and, by extension, the corrupt elite itself. Through the invisible force acquired during their purification and flowing through their bodies, les saignantes throw waves of energy at the minister, weakening him. However, they also continue to roam the streets of poor neighborhoods to escape state-sponsored forces. The exploration of the various political and socioeconomic strata where corruption, decadence, and death lurk demands collective healing, which is thematically central to the film. Although the film does not feature a final victory over the ruling class, it succeeds in showing the positive forces of secret societies, which are then co-opted for indirect political goals.

As a politically and socially engaged film, *Les Saignantes* restages the role that the Mevoungou could play in the current sociopolitical climate. However, by mixing different filmic genres, the film subverts heteronormative gender dynamics as well as form. Although *Les Saignantes* resists being pigeonholed, critics and scholars—including Kenneth Harrow (2005), Olivier Barlet (2007), and Olivier Tchouaffe (2006)—have compellingly analyzed the film as a subversion of genre and gender dynamics. For instance, analyzing its unconventional lighting and editing, French critic Barlet compares it to an abstract painting— more fantastical than realistic.

Bekolo achieves his political critique through the storyline and by structuring the film around six interstitial title cards. These strategically placed title cards reflect different film genres—action, detective, pornography, romance, mystery, and science fiction. Shaped like billboards and displaying rhetorical questions, the cards effect a visual rude awakening of the viewer's social awareness, bestowing on her social and political responsibility. For example, the first billboard-like title card asks, "How can you make an anticipation film in a country that has no future?" Another one poses the question, "How can you make a horror film in a place where death is a party?" Truly, death permeates the film, and several scenes use conventions of the horror genre. Because they are popular symbols, billboards not only emphasize the political urgency of the rhetorical questions but also allow the reader to momentarily step away from the macabre and violent atmosphere of the film. In other words, they function as markers that help redirect the viewer's attention. The final card at the close of the movie is the most telling and the most action-oriented: "How can you watch

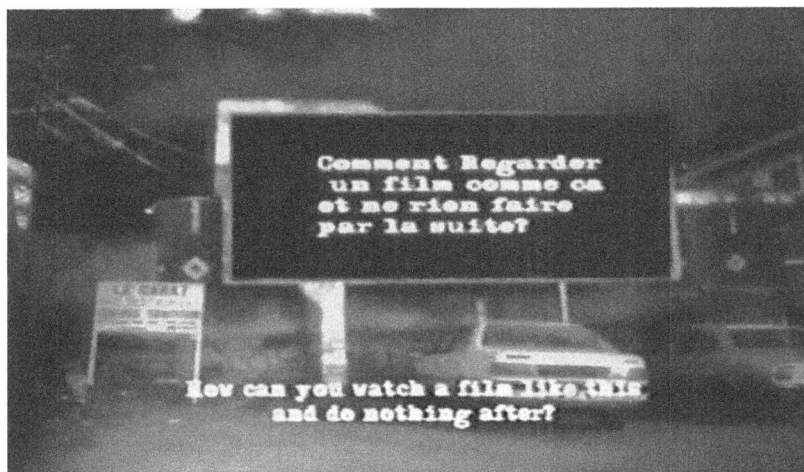

FIGURE 5.1 The last title card and frame from *Les Saignantes* (2009).
Courtesy of Jean-Pierre Bekolo, SARL.

a film like this and do nothing after?" (figure 5.1).[3] With this last title card, it becomes obvious that the film is a roughly disguised, scathing critique of the decadent and corrupt postcolonial state, marked by the absence of political agency.

Les Saignantes, which calls for change via a female secret society, was inspired by an actual event. In an interview in 2005, Bekolo bitterly explained how he discovered the intimate connection between an unsanitized, unpurified female sexuality, often reduced to transactional sex, and the decadence of the ruling elite: "During a stay in Cameroon, I wanted to meet a minister but I didn't succeed. A young woman told me she could sort it out for me. I quickly understood it was a network and I could have met the whole government that way! These young women control the workings of the system and they have a certain power" (quoted in Barlet 2005).[4] In several interviews, Bekolo explained his choice of the female society and his attempts to reclaim a feminine modality of power in a space where some consider women's bodies as instrumental in the corruption of the state and where masculine symbols have failed. He frames women's bodies, albeit a sanitized version, as both the poison and the cure: "I had the idea that if I focused on women, I would really touch on very sensitive issues in society. I was trying to make a film about Cameroon, and so it was important to bring up the issue of women's relationship with men in power. That is a sensitive issue, and it would seem more interesting than if the central characters were to be boys. Also, *there is a connection between the idea of human corruption and girls*" (quoted in Adesokan 2008, 2, emphasis mine).

In an interview with Claire Schaffner (2009), Bekolo considers women as instruments to draw attention to questions that would otherwise remain without interest, adding, "Concerning the girls, they become interesting when they adopt an ambivalent behavior because everyone gets involved: traditions, the police. . . . *It will be less radical if the characters were male. The girls allow me to politicize my views of society*" (emphasis mine).[5] These comments reflect the kind of exploitation of women's incorporation of indigenous African religious nature that we saw in Côte d'Ivoire with Germain Coffi Gadeau's comment about his political party's co-optation of women's movements against the French colonial administration (see scene 3). While Bekolo powerfully contributes to displacing the emphasis on the disempowering images of women's bodies as cut, violated, and diseased, he falls into the equally problematic trap of showing them as healing and empowering in order to take them over. To illustrate, the film somewhat resembles the political and literary discourses that sought to advance decolonizing agendas or to critique the failures of the postcolonial country without any regard for the interests of women themselves, as Ella Shohat (1991) shows in her larger postcolonial feminist reflection. Bekolo's explanation, "It would seem more interesting than if the central characters were to be boys," turns women into metaphors as well as marketing instruments in his political critique of the postcolonial nation-state. This filmic co-optation and adaptation of a female secret society demands rethinking the nature of the women's agency, which is figuratively naked primarily because it is ever unfolding and can accommodate self-determination, instrumentalization, instinctual response, backlash, and triumph. Hence, images of the Mevoungou are placed somewhere between the triumphant reading of the secret society as empowering and a view framing it as backward and holding back the march of the country toward a putative and undefined progress.

Bekolo's intellectual and professional connections constitute major influences on his filmmaking and the subject matters he tackles.[6] Questions related to the body, cannibalism, and women's sexuality, so pervasive in *Les Saignantes*, may reflect the influence of Bekolo collaborator Claire Denis. The French feminist film director grew up in Cameroon and collaborated with Bekolo on *Quartier Mozart* (1992). Her interest in the body and cannibalism is visible in movies such as *Beau travail* (1999), *Trouble Every Day* (2001), and *The Intruder* (2004). In addition to the Denis influence, it is safe to say that Bekolo's focus on the body and its disintegration and cannibalism stems from the postcolonial culture's fascination with the body and excessive consumption, which is best known in Francophone West African popular parlance as *la mangecracie* or *la politique du ventre* (the politics of the belly) (Bayart 1989). Excessive consumption permeates every sphere of the postcolony, to the point of becoming a hallmark of political ruling classes.

Like Sembène Ousmane in his filmmaking, Bekolo sees in cinema a means for social change, but unlike him, Bekolo rejects the documentary, the mimetic transposition of reality into the screen. Bekolo seeks to achieve consciousness-raising with thought-provoking and at times disturbing films.[7] After *Quartier Mozart*, Bekolo voiced his preferences for the extremes and the allegorical, which, he argues, stimulate imagination: "The vision of a people must not be reduced to its reality. This is sometimes the tendency of African films that have been made so far" (quoted in Verschueren, Tapsoba, and Maïga 1992, 16). However, transcending conventional boundaries did not come without challenges; the Cameroonian Film Commission rated *Les Saignantes* NC-17 and imposed a short-lived ban on it. The film was banned on charges of pornography, and the commission demanded that the director delete sections considered indecent. In reality, the ruling political class was embarrassed about how the film had exposed its weaknesses. The incident of the ban demonstrates political permeability in Cameroon, where supposedly independent commissions actually receive orders from the ruling party. The incriminating scenes disturb the decadent comfort of the male ruling class by metaphorically and literally baring its failures and emasculating it, using a female secret society.[8]

The Mevoungou and Anthropology

The recovery, adaptation, and larger circulation of an extinct women-centered society and ritual address the social and political intervention of the film. Because German missionaries abolished the Mevoungou in the nineteenth century, Bekolo, himself a Beti, did not hear about the society or the ritual until he discovered it through *Le tombeau du soleil* (1986) by French anthropologist Philippe Laburthe-Tolra, who taught at Yaoundé University and extensively studied the Beti.[9] The discovery of the ritual via a French anthropologist suggests the frequent nonlinearity in the perpetuation and transmission of cultures. Choosing to use the Mevoungou ritual allowed the director to reclaim an aspect of his culture that colonizers denied him. Thus, Bekolo meets and reclaims the ritual without the stigma of backwardness that some critics have attached to genital rituals of purification and cursing.

French and Cameroonian anthropologists—including Jean-Pierre Ombolo, Charles Gueboguo, Jeanne-Françoise Vincent, Philippe Laburthe-Tolra, and Marie-Paule Bochet de Thé—have published on the Mevoungou since the 1950s. Vincent's ethnography *Traditions et transition: Entretiens avec des femmes*

Beti du Sud-Cameroun (1976) provides an overview of the ritual based on seventeen interviews she conducted with Beti women in 1967 and 1971. Because the first German missionaries abolished the practice, only three of the interviewees who described the ritual, mostly in their sixties, had actually participated in it. Most of the accounts were therefore based not on firsthand experiences but on reports. Consequently, interviewees differed on the significance of the ritual, to the point of contradicting one another. One interviewee thought the ritual and its festivities were a way for women to eat meat-based meals, as men regularly enjoyed those meals at the expense of women and children. Others defined it as a ritual of purification, initiation, protection against evil eyes, celebration of femininity, and resistance against male domination.

According to Beti cosmogony, the Mevoungou group is built around the cult of the *evu*, the exterior physical site of the gland embodying good and evil that is located in women's uteruses. When activated, the gland is capable of giving life or taking it. The location of the gland explains why the power of the Mevoungou comes from the clitoris, considered to be the exterior physical site of this gland. The leader of the society, the Mother of Mevoungou—who is past menopause and no longer engages in heterosexual intercourse—is chosen based on the largeness of her evu (her clitoris), which is believed to symbolize the extent of her power and fertility. This criterion led Laburthe-Tolra to observe that the Mevoungou ritual is a "celebration of the clitoris and of feminine power" (1985, 234).[10] Adoration of the clitoris explains why, in Beti culture, its excision is the ultimate punishment for adulterous women.

Although the Mevoungou society is forbidden to men, the entire Beti community of men and women reaps its benefits. Its rituals, performed at the request of both men and women, aim to settle cases of thievery or witchcraft, to resolve personal crises, and to bring prosperity during times of drought and calamity in the community. The ritual is divided into two ceremonies (public and private); the public one reunites the entire village, whereas the secret one is exclusively attended by the initiated and the *mvon Mevoungou*, candidates for the society. The secret ceremony starts with the Mother of Mevoungou undressing and inviting participants to undress before she swears them to secrecy. Later, women invoke a ceremonial package composed of roots, ashes, medicinal leaves, and centipedes. The composition of the package, known as *mbom Mevoungou*, reinforces the belief in women's connection to nature. After the transmission of the power of the clitoris to the package, women rub, massage, admire, tickle, and stretch out the clitoris to give it the allure of a virile organ (Laburthe-Tolra 1985, 238).[11] Participants then mimic intercourse, older women assuming the conventional masculine roles and the younger ones taking the feminine. Throughout the ritual, women make fun of male genitalia, degrading them while celebrating female genitalia. Then, the women burn the package and divide its ash into three portions. The first portion is buried with a centipede in front of the hut of the organizer, the second is made into a package to represent the Mevoungou, and the third is sprinkled on rooftops and around the village. The rest of the ceremony consists of transmitting the power of the package to the clitoris of the Mother of Mevoungou and to those of other women. If the Mevoungou ritual is performed to unmask wrongdoers, the village expects them to confess, or else the villagers will experience an outbreak of diseases ranging from swelling to various skin conditions.

Overall, the ritual consists of establishing a communion among the different entities: the package, the clitoris of the Mother of Mevoungou, and the clitorises of other participants. Whereas modernity and its attendant institutions pathologize women's bodies, and existing ethnographies and anthropological literature wage a battle over whether the ritual was sexual or not, Beti cosmology reportedly venerated the ritual for its spiritual and healing dimensions.[12]

It is intriguing that those who desexualize the ritual happen to be men, even though the ritual is forbidden to them. I argue that the desexualization transcends the cultural battle between French feminist anthropologists and Beti intellectuals to enter a gendered landscape. Although the research necessary to argue the connection between rubbing the clitoris and sexual pleasures lies beyond the scope of this scene, I would say that by desexualizing the ritual, male observers discursively deny women who performed it the possibility of experiencing physical pleasures. Women's enjoyment of sexual pleasure during the ritual should not contradict or annihilate its "spiritual" or even patriarchal aspects. To ignore completely possible pleasures associated with the ritual is to "hijack" Beti women's bodies. Whatever the purpose of massaging clitorises— enjoyment or purification, or very likely both—Bekolo's film imagines and appropriates for sociopolitical purposes the constructive nature of the Beti female religious society.

Dystopia in the Making

As a fictionalized record of the challenges facing the unnamed country, *Les Saignantes* visually narrates the making of a futuristic "deathscape" in which degeneration and predation turn people into zombies and the politically dead in a process of political annihilation, which Achille Mbembe has called "mutual zombification" (1992, 4). The manufacturing of naked lives, entities that are exposed to death with impunity, becomes the normal paradigm of the governmental state. Naked life, not animal life, dwells in the dystopia that Bekolo paints. This differentiation draws on Agamben's (1998) conceptualization of naked life, which emerges because of sovereign violence from the distinction between *bios* and *zoe*. The monstrosity of Bekolo's imaginary country is that human beings are exhibiting animal-like behaviors, being reduced to naked lives.

The cab driver taking Chouchou to Majolie's apartment after the SGCC's death animalizes her when he threatens to rape her and predicts that she will howl like a dog. The taxi driver himself does not escape animalization: his debilitating stammering renders his speech incoherent, read *inhuman*. Additionally, the butcher to whom the SGCC's body was delivered for dismemberment is growling like a dog. As if stripped of his linguistic abilities and possibly of his political agency, the butcher's preferred mode of communication is growling when he approves and wielding the rusty chainsaw when he disapproves. With these disturbing images, unusual in West African cinema, most men act as if stripped of a fundamental human attribute: speech. From an Arendtian perspective in *The Human Condition* (1959), action is most centrally connected to speech (words) and to agency. Unlike labor or work, speech fundamentally enables the speaker to disclose her identity, and so speech becomes the sine qua non condition in the ascription of agency. Therefore, one rightly wonders about the possibility of reclaiming sociopolitical agency in a space designed to stifle speech and to engineer political death.

The macabre atmosphere of the movie enwraps and confounds everything, and one justifiably wonders if the society of 2025 is not the apocalyptic vision of an annihilated state. The voice-over, more informed, axiomatically claims: "We were already dead." Later, in a rhetorical question, she clarifies: "How does one recognize Mevoungou from others in this country where it's impossible to separate the living from the dead?" Indeed, in the dystopian society, it is impossible to distinguish the living from the dead because, just like the living often do, the dead also vomit, creating an obscene and senseless world. For instance, even dead, the SGCC still reacts to the outside world by vomiting in Majolie's face, thereby refusing to be confined to the state of silence. At first, the vomit causes Majolie feelings of revulsion. For a moment, Majolie believes in the resurrection of the body as the viewer searches for the meaning of an act that probably has no meaning except to materialize meaninglessness in the postcolony.

With regard to the banality of death and as a reminder, one title card asks us, "How can you make a horror film in a place where death is a party?" The characters even coin the expression "DGPs" (*Deuils de grandes personnalités*), translated as "WIPs" (Wakes of Important Personalities), to underscore the festivities accompanying death. Funerals and WIPs have turned from mournful gatherings to joyful celebrations where drinks and food are plentiful and where the poverty-stricken feast. The confusion of predation and production imitates the confusion of death and life. These are typical representations of paradoxes of the postcolonial space: rich in natural mineral resources, yet poverty-stricken and heavily indebted.

As the viewer navigates the litanies of monstrosities, the title cards pull her from what seems like a long nightmare. At the beginning, the film enlists her participation in a sexual adventure with an almost naked heterosexual couple. But the scene turns out to be about death. The low-key lighting combined with diffuse shadows and pools of light, marked by the continued movement of openness and closure of perspectives, is confusing, especially when death is a party and love is death. The assaultive combination of red, orange, green, and blue as dominant colors is all but reassuring. The sense of a thriller or mystery genre is reinforced with the blurry quality of the lighting.

Shot from below, Majolie literally and metaphorically looks down on the SGCC. After a moment, she springs on him and jumps away. Majolie's position in the shot and her gyrating, engulfing, and virile hips take the obedient older male organ from the terrain of the superficially pleasurable into that of war, submission, and death.[13] Diverging from conventional readings of this scene as sexual, I propose that it is a purification ritual wherein the powerful male authority figure is cursed and sacrificed through Majolie's genitalia. The dynamic of teasing, so manifest in the scene, mimics the director's play with received norms and comfortable expectations. The integration of dance/celebration and shooting/death, two seemingly contradictory phenomena, blurs the boundaries between life and death. The combination also turns traumatic events into regenerating ones, reflecting the creativity of the postcolony, which Mbembe calls "afropolitanism" (2013).

The erotic/death scene pushes the boundaries of African cinema further by showing close-ups of Majolie's pubic hair and her breasts. In African cinema, it is unusual to bare the female character's naked lower body. According to Alexie Tcheuyap, African films portray and often allude to sex, but they usually show no sexual acts on the screen: "*Trompe-l'oeil* and veiling strategies are used by filmmakers to move sex away from the screen. Sex is represented in films, but implicitly rather than openly. It is present in the story, but absent from the plot. In this context, ellipsis plays a fundamental role" (2005, 145). Typically, the lower sections are framed and/or hidden through strategies of substitution, including long shots, dark screens, jump cuts, and ellipses. Contrary to other directors, Bekolo chooses to expose breasts and pubic hair. With exposed genitalia, the scene is provocative but also titillating in its withholding of details. The absence of dialogue preceding and during the scene contributes to the suspense level, with the voyeuristic handheld camera, generally used in documentary-style realistic filmmaking, compounding the anxiety.

The spatial dilapidation—war-torn landscape—reinforces the loss of markers in the dystopia. In a sort of unending nightmare, smoke incessantly covers the sky and envelops everything in its path. After the death of the SGCC, Majolie calls her friend Chouchou for help; during the latter's cab ride, the viewer has the opportunity to assess the extent of the spatial decrepitude. The slow motion and the long shots show the streets with unsightly and uneven pavements, holes, smoke, and moving and indistinct human figures. The discomfort-generating sight is eerily reminiscent of the images of African cities as routinely depicted on television screens by international news channels. The pervasive foul odors of dead dogs and, probably, human corpses that Chouchou notices are reminiscent of a wasteland or of a war-devastated area.

The voice-over, however, relieves the anxiety: "The country could not continue like that without a future. It had to change." Truly, it has to change through a purification ritual because "there was no place for despair." The cleansing process involves emasculation and theophagy. Rather than men and masculine modes of power and production, which have failed, women and traditionally feminine forms of power take center stage. Women's occult forces will liberate the country from the claustrophobic space and purify it. Thus, the film moves beyond normative frameworks, instead proposing a degenitalized ritual to replace what was originally a female genital ritual as a signifying practice to reach collective healing and freedom. The Mevoungou ritual is thus adapted and co-opted for communal purposes, just as in precolonial and colonial times among the Beti. However, an appropriation may not primarily serve women because of the prevailing gender configurations in the postcolonial state. If in precolonial and colonial Beti societies, women's secret institutions and rituals were used to serve communal goals, women equally enjoyed certain prerogatives, which they lack in postcolonial cities with their differing power brokers.

Genital Purification

The necessity for performing the ritual is unequivocal, as the offscreen voice gravely invites and coerces after the shocking first scene: "Mevoungou has fallen on us like a bad dream. Now this country has a chance to escape from the darkness." "Mevoungou was inviting us to join the dance." Mevoungou offers the country a way to end repressive violence. The use of the communal plural is an invitation to the viewer to join the community of men and women who need purification through women's creative and healing powers, even if the film at times desexualizes the ritual.

The Mevoungou ritual is central to the film, controlling and guiding all the characters. It is introduced after a blackout, following the genital scene between Majolie and the SGCC, a moment of heightened expectation and confusion: "Mevoungou is neither a living being nor a thing. Mevoungou is not a place much less a moment. Mevoungou is neither a desire nor a state of mind. Because Mevoungou is something we see, we live and experience but cannot quite define. We don't decide to see Mevoungou. Mevoungou appears to you. Mevoungou invites itself."

Like a dormant entity ready to intervene in challenging circumstances, the spirit takes possession of les saignantes for the ritual of purification. The adapted Mevoungou ritual is defined in intangible terms. It is no longer time- and place-specific; from its original setting, the village, it is now relocated in the postcolonial city, where corruption reeks and women lose their bearings. As a metaphor of African mothers' genitalia, the Mevoungou ritual is invoked and performed to cure the ills of the found(l)ing fathers.

Nighttime, in which the entire film is shot, is the conventional time to perform the secret Mevoungou ritual. Through the lack of light, the entire movie creates a paranormal and gripping atmosphere that unsettles the viewer. The thriller-like mood, created through the simultaneous use of diffuse shadows and pools of light, thrusts the viewer into an atmosphere of instability and chaos. The omniscient narrator doubles the offscreen voice, a device that forces the viewer to consider the larger implications of the issues tackled in the film. The narrator informs the viewer of plot development while providing a haunting mood.

To perform the purification, the women undergo several rituals of cleansing. The first of these is a ritual bath that Majolie takes after the SGCC's death. She impulsively and vigorously scrubs her body, and after the physical cleansing, she undergoes an internal purging. Still possessed, Majolie walks into a bar and grabs a bottle of hard liquor, "a sacred fluid" (Akyeampong 1996, 164), from which she drinks. With the physiological mind-altering effects of alcohol, believed to facilitate the interaction between the physical world and spiritual world, Majolie accesses the spiritual world, the higher truth (Akyeampong 1996). Her rite of passage continues through the uncontrollable consumption of alcohol, after which she urinates and simultaneously experiences orgasmic spasms outside the bar, a foot away from other bar patrons. The background shot in the urination scene suggests the publicness of purification, which was

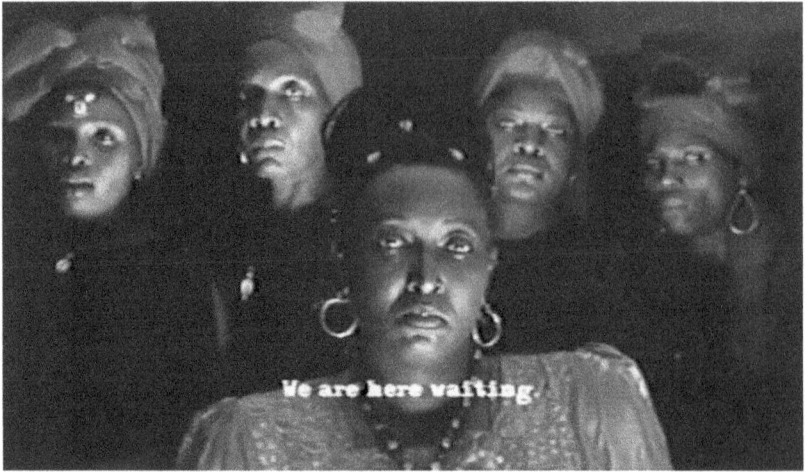

FIGURE 5.2 The initiated, in a film still from *Les Saignantes* (2009).
Courtesy of Jean-Pierre Bekolo, SARL.

invested with transformational powers, as Majolie groans and seems to be push-
ing out of herself the creeping stains of pervasive corruption. This first stage of
the purification gives Majolie and her friend the strength to dismember and
dispose of the SGCC's body.

With visual effects, the initiated members of the Mevoungou, represented
here by five elderly women in red and blue headscarves with serious and authoritative demeanors (figure 5.2), guide and protect les saignantes. In the middle of
the film, when the police suspect the protagonists' subversive activities, Inspec-
tor Rokko goes to Chouchou's mother's house to arrest them. In his search, four
women who are invisible to him caress him; they lift his gun and move him in
an instant from the house to a hospital.

During the process, when the initiated attempt to touch and assess Chouchou's clitoris to empower her, she refuses their advances.[14] Given the challenge
of symbolically feeding food to and stretching the clitoris on screen, the film
avoids the conventional clitoris-centered ritual and instead features a dance to
Brenda Fassie's "Vuli Ndlela" (2002) as an alternative method of purification
(figures 5.3 and 5.4). *Vulindlela*, a Zulu word that means "to open the way," is
a very suitable incantation for a transformational session. Through the elaborate frenzied dance scene, rather than a clitoris-centric ritual, Majolie and
Chouchou change outfits several times and embody several types of women;
they come out transformed and equipped with what the offscreen voice calls

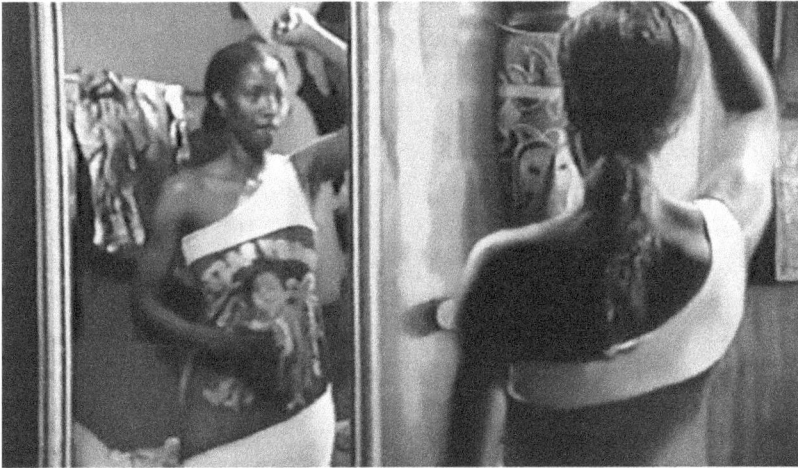

FIGURE 5.3 Dancing to Brenda Fassie's "Vuli Ndlela," in a film still from *Les Saignantes* (2009). Courtesy of Jean-Pierre Bekolo, SARL.

FIGURE 5.4 Dancing to Brenda Fassie's "Vuli Ndlela," in a film still from *Les Saignantes* (2009). Courtesy of Jean-Pierre Bekolo, SARL.

their "integrity." At a dizzying pace created by the doubling and tripling of shots, Majolie and Chouchou metamorphose from sexy to demure, from young to old, and from trashy to respectable. The doubling and tripling effects condense time and space and reject the comfortable position of sameness and immobility by creating an uncanny sense of both estrangement and familiarity. Here, dance movements take the women into the spiritual realm, where they embody several beings and live several lives, continuously creating and evacuating identities. Consistent with anthropological descriptions of the ritual, from the first scene to the last, the women's dancing resembles ritualistic movements. Although they may appear terrifying and zombie-like, they are nonetheless conforming to the demands of the ritual. Such movements are "not used in the representational mode, that is, to signify but instead to create psychological states. These elements are the inevitable elixir, as it were, for the performers' transition into liminal states of trance and religious performance" (Conteh-Morgan 1994, 31).

The power invested in songs and dances becomes obvious when the offscreen voice announces, "Our Mevoungou had rediscovered its integrity. It could no longer accept the slightest insult. We were ready for the final phase." Through the dancing frenzy, the protagonists recover their agency to be constructive members of their society, to come out as newly born women, ready to deliver the last blow to the enemy. Following the purification, they attend the SGCC's funeral and subsequently disarm the minister.

The scene of metamorphosis is made even more compelling with Bekolo's use of the mirror. While dancing, Majolie and Chouchou gaze into a mirror and watch their appearances change. The specular image provides the protagonists a triumphant turning point in their identity formation. They reach their completeness and integrity, unlike in the previous stage, which was characterized by "uncoordination" and fragmentation. Fragmentation—of both the SGCC's body and of the nation-state—is a pervasive theme in the movie.

To purify the nation-state, the Mevoungou ritual sacrifices, emasculates, and engages in theophagy. The ritual often included the sacrifice of a chicken or goat whose blood was used for purification (J. Vincent 1976, 20). In the film, the sacrifice of the SGCC becomes the sine qua non condition for a more prosperous state. The title of the film, *Les Saignantes*, means in popular French "cruel women," or those predisposed to inflict pain. That name resonates with the women's mission to purify the diseased country through bloodletting. Their mission suggests that the appropriate English translation of the title would be "the bloodletters" rather than "the bloodettes" or "the bloodiest."

As transformed women, Majolie and Chouchou threaten authorities to such a degree that the police officer finds it necessary to inform the minister of the interior of the women's imminent plot against the state: "I've [*sic*] to warn the minister of state that a danger is menacing the republic." The supposed threat suggests that women possess the power to undo the corrupt postcolonial state. After all, they did kill, decapitate, and reconstitute the body of the SGCC. In fact, toward the end of the film, women's genital purification proves deadly for the minister and, by extension, the ruling body. In the fierce battle, they tap into their ritual energies, release supernatural forces, and lift him off the ground, presumably killing him. Women's orgasmic release means the temporary demise of evil in the film. In that sense, their purifying ritual cannot be separated from their political strategies because it offers the potential to undo the postcolonial nation-state, as M. Jacqui Alexander (2005) has formulated in a Caribbean context.

The scene of Chouchou and Majolie's degradation of the SGCC's testicles is another ritual element that is consistent with anthropological descriptions of the Mevoungou ritual. After the protagonists have the butcher decapitate the body, they take a break from their macabre tasks and use the package representing the SGCC's testicles as their toy; Majolie asks, "The Secretary General of the Civil Cabinet's balls?" Toying with the SGCC's testicles indicates how young women dare disrespect old men, the supposed possessors of social and political powers. Insulting and degrading male genitalia are key aspects of the nightlong Mevoungou ritual. In that regard, Laburthe-Tolra (1985) reported that German anthropologist Gunther Tessmann found the Mevoungou songs so outrageous and degrading that he could not report them. Kicking the SGCC's balls, insulting his manhood, constitutes his ultimate defeat as a man in a system where men's genitals are armed with the power of a weapon. Male anxiety over the loss of genitals reflects the fear of extinction of the self and the species. The genitals are not only crucial in reproductive functions; they also play a defining role in the power dynamic between men and women. Disappearance of male genitals, a recurrent allegory in African films, is synonymous with the collapse and failure of the state.[15] *Les Saignantes*' symbolic emasculation of the ruling class is in itself a painful and humiliating blow to the imagined patriarchy and dictatorship. Modern forms of power could not withstand the forces of a resurrected female secret society of a religious nature.

Most often, the film portrays the protagonists as vampire-like, operating at night. They have prominently red lips and glistening and bloodshot eyes— created by colored contact lenses. Unlike the conventional readings of vampires as destructive, here they represent positive entities on a mission to save the country. The revealing poster of the film emphasizes the women's bloodsucking

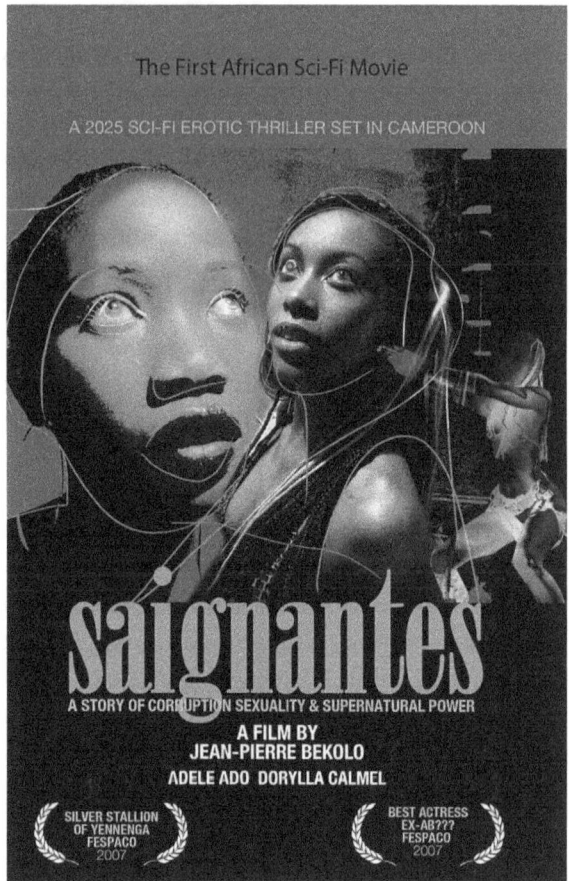

The First African Sci-Fi Movie

A 2025 SCI-FI EROTIC THRILLER SET IN CAMEROON

saignantes

A STORY OF CORRUPTION SEXUALITY & SUPERNATURAL POWER

A FILM BY
JEAN-PIERRE BEKOLO
ADELE ADO DORYLLA CALMEL

SILVER STALLION
OF YENNENGA
FESPACO
2007

BEST ACTRESS
EX-AB???
FESPACO
2007

FIGURE 5.5 *Les Saignantes* poster (2009). Courtesy of Jean-Pierre Bekolo, SARL.

qualities, showing the initiated as devourers of the corrupt ruling elite, engaged in an economy of emasculation and consumption (figure 5.5).

Although the decapitation and the emasculation scenes are unsettling and create a sense of malaise for the viewer, they also empower her by reversing the social hierarchy, potentially turning the SGCC's body into meat for the people's consumption. This consumption suggests the possibility of consuming the entire decayed system. The potential ingestion of the SGCC's body becomes a practice of ritual cannibalism. In the film, we see a shift in the structure of perception of cannibalism, from its reading in colonialist writings as a sign of "absolute difference" between the natives and the colonialists (Greenblatt [1991] 2008, 135; Stam 1989) to a form of domination in the postcolony. That form of domination shows the act of eating as a negotiation of power dynamics between the eater and the eaten, as Maggie Kilgour (1990) argues in a different

FIGURE 5.6 At the butcher shop, in a film still from *Les Saignantes* (2009). Courtesy of Jean-Pierre Bekolo, SARL.

context. Indeed, in the discursive economy of colonialism, cannibalism, real or imaginary, was constructed as a sign of the natives' backwardness and consequently was used to justify the colonizing mission. The film's reversal of the trope of cannibalism contributes to its overall message of doing away with reified practices.

The film becomes even more compelling as it moves from cannibalism to theophagy, a higher level of subversion. As in Mbembe's postcolony, here the "worshippers" devour the "gods" (2001, 112). This kind of ritual cannibalism is central to Christianity: the wine and bread that signify the body and blood of Christ. In Catholic belief, Mass involves transubstantiation, presumably the actual body and blood of Christ. The scene at the butcher shop (figure 5.6) refers to the power that the Mevoungou ritual has harnessed to purify the state, which threatens the authorities so much that the police officer informs the minister of homeland security and, by extension, the president of the imminent threat that women are plotting against their rule. Bekolo's version of theophagy seems necessary to turn the country into a self-sufficient one, especially if we consider that the corrupt elite metaphorically and literally feed on the resources of the people; this autonomous country is capable of flourishing without the handouts of international financial institutions. The supposed or real threat to the existing state speaks to women's co-opted contribution to the reconfiguration of the new state. By choosing female characters as active agents in the administration of such a cure, the film, albeit problematically, reclaims

an indigenous female religious institution that is endowed with both beneficent and maleficent powers.

With these complex representational politics, *Les Saignantes* frames wom-
en's bodies as both the poison and the cure, a sort of Derridian *pharmakon*,
an aspect that scholars have not noted in previous scholarship on the film. Although that framing of female bodies seems to move beyond the dualistic construction of women's bodies as pathological versus purifying, the film, however,
falls into another binary: the opposition between the collective purifying power
of the female society and the individualized and transactional uses of genitalia
by the titular characters before the intervention of the Mevoungou ritual.

Use of the precolonial female customary ritual geared toward communal
survival is remarkable because it shows that there is more to images of African
women's bodies than pathology. Thus, this filmic interpretation contradicts
conventional African as well as Western feminists' views of "tradition" as the
locus of women's subordination and modernity as the path to their liberation
(Arnfred 2002). However, *Les Saignantes* also engages in the bifurcated logic
of "bad" versus "good," which has served neither African women nor African
feminist theory well.

Visualizing Naked Agency

Although the choice of cinema as a vehicle of social consciousness seems indisputable, one can question the suitability of *Les Saignantes* for a general audience. The disappearance of the clitoris from the purification ritual is a double-
edged undertaking. Bekolo moves away from the clitoris-centered ritual and
instead makes the choice of a performative cliché of female sexuality—trying
on clothes—rather than praising the clitoris. In that context, the movie degeni-
talizes what had been, at least physically, an exclusively female genital ritual. I
argue that the disappearance of the clitoris answers the director's need to get
around the challenges of representing the clitoris on the screen, thus avoiding
divulging the secrecy around the ritual or violating women's bodies, or both.
Perhaps Bekolo is echoing his fellow film director Anne-Laure Folly, who argues, "African culture is more discreet, less externalised, more modest and re-
strained. Every disclosure of the matter is a violation of the matter. In this sense,
cinema is a transgression" (quoted in Barlet 2000, 104).

Les Saignantes is part of an emerging set of cultural products that grapple
with the proliferation of unconventional nakedness. Given the sensational
nature of the gesture combined with the speed at which images travel in our

screen-saturated era, these artistic artifacts contribute to the co-optation of women's gestures both locally and internationally. These images of African nakedness by both self-defined Africans and others compound the already calcified images from centuries ago.[16] The images contribute, often unwittingly, to "other" the continent, as if it were "behind the existing world, out of the world" (Mbembe 2001, 173). To navigate possible accusations of impropriety, exploitation, and voyeurism while giving an account of the increasing importance that defiant disrobing holds within postcolonial biopolitical circumstances, visual artists and filmmakers employ comic strips, illustrations, paintings, and other stylized animations. Two films that have successfully adopted that strategy include *Pray the Devil Back to Hell* (Disney 2008) and *The Naked Option: A Last Resort* (Schermerhorn 2010). The first documentary tells of Liberian women's contribution to end the second Liberian civil war in 2003 and the second chronicles how mature women of the Niger Delta expressed their grievances against unrestrained rule and mismanagement of natural resources. Unlike *Uku Hamba 'Ze (To Walk Naked)*, analyzed in scene 2, both films use stylized illustrations in what I have called elsewhere "the cinematic language of naked protest" (Diabate 2017).

In 2004 anthropologists/sociologists Leigh Brownhill and Terisa E. Turner began a collaboration with Seth Tobocman to create a nine-page comic book called "Nakedness and Power" (Tobocman, Brownhill, and Turner 2006, 204–11). It features the 1991 Freedom Corner's protest with Wangari Maathai, the internationally renowned Kenyan environmental political activist and 2004 Nobel Laureate, as well as other women-centered movements across the continent. Maathai joined a group of elderly rural women in hunger strikes and public demonstrations to demand the release from jail of their political-prisoner sons. However, the direct action of hunger strikes turned into naked agency when the police attacked the women with clubs and tear gas.

These sophisticated and convoluted ways of insurgent self-exposure are both poignant and mundane. The poignancy speaks to the ethical conundrums the artists grapple with. Yet their creations also unwittingly dilute the impact of the angry women's gestures. It is in this way that adaptation of rituals of cursing on the global scene—whether through social media, visual art, or film—is inherently a challenging task. Adaptation is more likely to fail than succeed given that it responds unquestioningly to the call for transparency, which leads to the voracious need of consumers for the next enthralling image or news report.[17]

Social-media materials (discussed in scene 7), other kinds of visual representation (such as Onobrakpeya's artwork discussed in the introduction), and film contribute to a deeper account of naked agency. They not only give us access to the interiority of the women but also contribute to the dissemination and archiving of forms of purification and cursing, often serving to counter a triumphant reading of the women's agency and defiance. They oppose a triumphant reading in providing empowering and counterproductive details about the women's feelings and rationales and those of their targets and other participants in the drama of disrobing.

With film and media democratization, genital purification becomes "modernized" as groups of women mobilize social media and filmmakers marshal filmic strategies to further or complement the shaming or cursing effects of women's bodies and institutions. Feature films, Facebook posts and pages, YouTube videos, and tweets allow the vulnerable yet still powerful women, those who understand themselves as the moral guardians of their countries, to make a mark, although the mark may be short-lived or may even backfire. The openness that characterizes nakedness is crucial to this conceptualization of insurgent nakedness.

This formulation of openness as both enabling and disallowing moves beyond the bifurcated logic of victimhood versus sovereignty to suggest that women's agency, especially in relation to naked bodies used collectively, is ever-unfolding and empowering yet precarious. As women publicly contribute to the restructuring of their countries in the neoliberal biopolitical frame, they and their modes of contestation and retribution often are deritualized in a secularizing process. The most frequent of all reactions, secularization, demonstrates the unfinished business of the postcolonial nation-state regarding its indigenous religious practices, as I show in the next scene. Yet to consider secularization as only constraining is to disregard the ways in which it enables female protesters, including South African students of the Fees Must Fall movement, who are not banking on the sanctity of their bodies or their status as moral guardians to inflict shame.

Section III

Repression

Secularizing Genital Cursing
and Rhetorical Backlash

In the face of this [desecration], we will remain naked in front of the resident until we know why they did what they did. . . . To come and see us naked, that is an action that we refuse to tolerate. —**Yvonne Kouamé** (quoted in Jean Paul Loukou, "Yamoussoukro")

The Delegation of Women's Affairs do not encourage women's associations that operate as secret societies such as the Takumbeng of Nkwen and the Fombuen. . . . We encourage development associations. The practices of these secret societies, which operate as associations, are disgusting and expose women's privacy (walking naked on the streets). —**The Provincial Delegate for Women's Affairs** (quoted in Charles C. Fonchingong, Emmanuel Yenshu Vubo, and Maurice Ufon Beseng, 2008)

I define "secularizing genital cursing" as efforts by public officials to strip the gesture of its assumed religious connotations, especially its putatively harming and murderous implications. Although most events involving insurgent nakedness do not qualify as, or even pretend to be, genital cursing, secularizing seeks to turn such rituals or protests into a violation of decorum in order to make them available for legitimate state repression. Some of the repressive strategies in The Gambia, South Africa, Nigeria, Uganda, and Cameroon include legal threats, verbal attacks, physical brutality, arrests, and even murder. By implication, deritualization is for civic leaders a way to appropriate more power in order to secure the well-oiled march of the nation-state toward supposed political secularism with its promise of economic development and greater political freedoms. Given that secular governance supposedly shuns religiosity, indigenous religious practices and their spirits, of which genital cursing is a constitutive part, are thought to have no space within postcolonial biopolitical circumstances. In the deritualizing process emerges a set of figures that I call "translator-mediators." Against civic leaders' disparaging comments, these figures speak for the protesters/ritual participants in order to make legible their assumed goals. The mediators' intervention between the act and its reception energizes my account of the women's agency as precarious.

Secularizing effects are complex, both empowering and constraining women's agency. In fact, the women who defiantly disrobe are always already caught between multiple conflicting epistemic orders. The backlash against them reflects that conflict, and the ensuing debates in Africa constitute an explosive flashpoint around questions of indigenous religious practices and collective female political agency. More specifically, these rituals and the often contradictory responses that naked protest receives highlight unequivocally the unfinished business of the postcolony regarding its indigenous religious practices. Yet to consider secularization as only constraining is to disregard two factors: the violence of the targets may reflect their belief in the power of women's self-exposure and their attempts at exorcising it, and deritualizing the act benefits protesters who are not relying on the sanctity of their bodies in order to inflict shame. The incessant negotiation of power relations between these two poles can best be understood in terms of openness, of naked agency.

"Irreligious," "Anti-Islamic"

To deritualize genital cursing is to frame it as irreligious, as was the 2001 case of Gambian women. During the September 2001 Gambian presidential elections, a female society organized a mock burial of the opposition in conjunction with ritual nakedness. One evening, about thirty starkly naked mature women carried empty calabashes, cursed, swore, prayed, and denounced what they claimed was a distasteful ritual by the opposition to sacrifice a dog for election purposes. Lamin Dibba reported on the ritual in *The Independent* newspaper in Banjul, the country's capital (2001).[1] According to Dibba, early one morning, passersby discovered a dead dog, apparently killed and wrapped in white satin cloth with bottles of water placed next to it, near the Brikama Fire Brigade station, the location of the women's ritual. Hundreds of shocked onlookers watched the procession of clapping and shouting women stop near the Fire Brigade. Chanting in Jola, the language spoken by 10 percent of Gambians, the women prayed, "God punish the opposition for the evil act" (Dibba 2001). During the hour-long event, the ritual participants dug a hole, over which some sat while the rest continued to chant in anger. Lamin Dibba, the reporter, clarified that when he attempted to interview one of the women, he was flatly rebuffed, with the excuse that he was too young even to witness what they were doing. Following their "unusual and occultic practice [*sic*]," the participants left the location for an undisclosed part of town (Dibba 2001). In 2014, after an interview with the Jolas conducted on my behalf by Dr. Siga Jallow, a Gambian herself, Jallow revealed that the 2001 women's ritual is a Jola indigenous religious practice involving shrines and idols to ward off evil.[2]

By reportedly praying to what appears to be a Christian or Muslim God to punish the opposition, the women's invocation suggests the blending of an indigenous religion (and its reliance on the mystical forces in women's bodies) with the Christian or Muslim God. In this context, the Christian/Muslim God is not antithetical to other mystical entities but rather seems to complement them. Perhaps the women are aware of the inherent contradiction in their amalgamation of supposedly opposed religious orders and thus seek to transcend them in what Jean Comaroff (1985) in a different context has called "bricolage." Alternatively, perhaps the inclusion of the Christian/Muslim God speaks to the accommodating nature of most African religious practices, and their openness to novelty and change. The ritual participants' heterodoxy could also be an example of plural religious allegiances, although perhaps they are themselves unaware of possible contradictions.

With the city as the location of the ritual emerges a set of questions, including the legibility and effectiveness of cursing on the opposition, which is possibly foreign to the Jola cosmology that the women are mobilizing. Perhaps they intend motherhood and its functions as universally understandable, an assumption that arises from the shared fundamentals of biological functions. Yet the ritual problematizes several belief systems, most importantly those of numerous African communities, as explicated by Claude Lévi-Strauss's concept of the "shamanistic complex" ([1943] 1963, 179), according to which the effectiveness of curses such as these relies on the cooperation of the community. From Lévi-Strauss's perspective, to organize the ritual outside of its originary locale and in context with many potential bystanders and uninformed participants may be counterintuitive, ineffective, and illegible. Perhaps it is fair to assume that the mystical aspects of the curse are less important than the profound need for these ritual participants to expose their grievances to the world at large.

Despite what appears to be religious bricolage, and days after the news of the women's ritual became public, religious leaders from multiple denominations vehemently condemned the women and the rituals. Several imams, Islamic scholars, and a pastor strongly judged the ritual as "public indecency," "anti-Islamic," "backward," and "irreligious" (*The Independent* 2001). Abdoulie Ceesay, a renowned Islamic scholar, is quoted as labeling the protest "anti-Islamic, anti-society and anti-cultural." Sheikh Gibril Kujabi, the *mudi'r* (principal) of the Tallinding Islamic Institute, is paraphrased by a reporter as saying it was "profoundly reprehensible in the Islamic point of view." Using the language of morality, Sheikh Kujabi urged Gambians to express their grievances in "a more decent way" (all quoted in *The Independent* 2001). Imam Baba Leigh, one of the most powerful imams in The Gambia at the time, said that it was regrettable that this form of ritual was linked with Gambian politics. Unlike other religious leaders with their more vitriolic rhetoric, Pastor Edwund Martins of the New Fellowship International Church took a more tolerant, albeit patronizing, position regarding the ritual. He is paraphrased as describing the women's exposure of their "intimate parts" as "vile and repugnant" and a behavior that ignores "all religious and cultural restraints" (*The Independent* 2001).[3] However, Pastor Martins found the act forgivable because it was done out of ignorance. The leaders all called for the arrest and trial of the women, urging the government to clamp down on this event to dissuade its recurrence (*The Independent* 2001).[4]

The acerbic rhetoric of the religious leaders produces a climate of fear in which to participate politically in a certain way "is to risk being branded with the heinous and patronizing appellations" (Butler 2006, xix) of immorality and backwardness. In that sense, the women have to wage multiple battles since they must discount or embrace terms of the appellation, brave the stigma rising from the public domain, and, finally, seek reparation for the perceived wrong committed. The terms and the attack end up obfuscating the wrong that they attempted to right through their rituals.

The Gambia, a former British colony, is 90 percent Muslim and 8 percent Christian. The remaining 2 percent practice indigenous religions. Clearly, these condemnations index the religious leaders' rejection of and perhaps disdain for indigenous religious practices. Those familiar with debates in studies of African religions would agree that there are manifold definitions of African indigenous religions.[5] I follow the conventional definition offered by Jacob K. Olupona (2014) as being the religious beliefs held by Africans before Christian and Islamic colonizations. By their very nature, these religions refer to no single sacred texts, are plural, and are usually informed by one's ethnic identity and family heritage. In spite of their multiplicity and flexibility, a thread connecting them is the belief in the existence of invisible forces that influence the visible world. Despite the openness of these indigenous religious practices, they have been the object of attacks since the onset of colonization.

In the postcolonial era, the vilification-cum-deritualization of these collective rituals demonstrates the state's failure (willingly or not) to secure their constitutional and legal recognition. In many cases, the postcolonial nation-state inherited the colonial-era vilification campaigns of indigenous religious practices. The forms that such vilifications have taken differ slightly from those that criminalized the rituals a few decades earlier, and range from constitutional and juridical silence to outright arrest and incarceration of presumed members. Makau Mutua writes regarding state reaction to indigenous practices: "African constitutions and laws are generally either openly hostile to African religions and culture or they pretend that these kinds of religions do not exist. Such pretense is a tacit hope that African religions have either been eliminated or marginalized and so fundamentally delegitimized that they warrant no attention" (2002, 119).

To expand upon Mutua's explanation of the absence of normative recognition, I suggest a few possibilities, among which are the challenges of legislating practices that are inherently defined by their multiplicity and flexibility (no formal and written sacred texts and creeds), fluctuation between religion and culture, and freedom from clear designation and centralized leadership. Independently of the reasons behind the silence, indigenous religions have become "religious freedom misfits" (Hackett 2013), with increasing popular accusations against them of committing human rights violations and ritual and retaliatory killings (Gumbu 2010; O'Carroll 2015). This retaliation is supposedly because of the religions' numerous indescribable and unidentifiable taboos to which outsiders may fall victim.

"Savage and Brutal Africa-Style Practices"

The Gambian religious leaders' denigration of mature women's ritual disrobing is not an isolated incident. I found similar insults across the continent, in South Africa, Côte d'Ivoire, and Nigeria. In 2011 hundreds of Ivorian women symbolically exposed their menstrual cloths in Paris to curse American, French, and Ivorian politicians for the intervention of the French army in the 2011 postelection conflicts in Côte d'Ivoire. I discuss these events in detail in the context of epistemic silence in the next scene. Here, these events are useful to recall for the spiteful attacks to which the women have been subject. In "Opération 'Kodjo rouge' à Paris Comment le FPI se ridiculise" (Operation Kodjo Rouge in Paris: How the FPI Make a Fool of Themselves)—a March 27, 2012, opinion piece in Le Patriote that received readers' comments—Lacina Ouattara castigates the organizers of the rally for humiliating themselves and, by extension, their country. The weaknesses of the women's mode of protest, he argues, stem from their immigration to France and their "savage" form of protest. First, he points out what he considers ironic, that the adventurers will "flee" their country for a better future while purporting to protest in order to effect change therein. Second, Ouattara condemns the transfer by unsophisticated women in a civilized country of "savage and brutal Africa-style practices."[6] The illegibility of the Kodjo Rouge protest for a French audience, according to him, speaks to the protesters' inability to speak the language of law and democracy. The series of binaries that he sets up (barbaric vs. civilized, Africa vs. France, village vs. city, immigrants vs. host country, unsophisticated African women vs. democratic institutions, superstition vs. political rationality) muddles the complexities of a customary practice of prosecuting perceived injustice. The one-dimensional framework through which Ouattara analyzes these ritual cursings (the women's term) and their tactics reflects the monologic frame of reference that characterizes most debates on women, rituals, and contestation.

Readers' comments on Abidjan.net to Ouattara's vitriolic rhetoric reiterate the same monologic reference that the article set up.[7] Indeed, most comments focus on the immigration status of the protesters and the sexual aspect of their activism, chastising them for using their bodies as bargaining chips, which, in their opinion, is a salient emblem of the women's backwardness and sexual licentiousness. Of the fifteen comments, all of which condemn the women, seven are worth referencing for their offensive rhetoric and their ethnicized opinions. Three of these seven comments are explicitly said to be connected to the main university in Côte d'Ivoire, and only one commenter is identified as female. Several readers, deducing from the ethnicity of Laurent Gbagbo, a Bété, and of Kodjo Rouge organizers (Bétés), call all the women Bété and highlight the fact that the national imaginary considers the Bétés as shameless, a myth created by the French colonial history (Dozon 1985a, 1985b).[8] The blatant ethnicization of comments is manifest in the readers' names, which include those that sound as if they are from Gouro, Malinké, and Akan sources.[9] The ethnicization is in keeping with political alliances formed around ethnic and religious lines, a marked effect of colonization.

One reader points to the protesters' sexual looseness: "'low-life' and 'scorned' women who were born during shame's funeral wake. They give French people the opportunity to distinguish between good Ivorians, those who are proud of their bodies and respect French republican values, and bad ones, those whom these outmoded 'low-class hookers' represent" as quoted above (Ouattara 2012).[10] Another suggests that by donning what he calls "bikinis" and performing the eroticized Ivorian dance, the "Mapouka *dedja 'couleur sauce graine,*'" the women are displaying their protuberant bodily features in order to attract single

French men. The *Mapouka dedja* refers specifically to the pornographic version of the dance. Finally, a reader identified as affiliated with le Faculté de Droit (the Law School of L'Université d'Abidjan à Cocody) opines, "It is another form of savagery to wear such an attire in a country not one's own" (Ouattara 2012).[11] Again, the binary opposition of the civilized and the barbaric, the ethniciza-tion of this mode of political contestation, and the focus on the protesters' lack of modesty and immigration status fail to account for women's organizational abilities and political awareness.

"Anarchy, Chaos, Mayhem," "Disgusting"

What is more surprising in this vilification trend is the participation of academ-ics. In South Africa, the chastisement is especially acute. Speaking on the 2016 naked protest by the African National Congress (ANC) youth, political analyst Somadoda Fikeni found the strategy to index "serious social challenges includ-ing the manner of engagement" (quoted in Moatshe 2016). Eight years earlier, Makhudu Sefara (2008) scathingly criticized naked protests in South Africa as political barbarity: "anarchy, chaos, mayhem—the very step-cousins of barbar-ity." In Nigeria, Femi Ajayi (2010) called a group of women who threatened to go naked if Senator Gbemisola Saraki refused to run for the 2011 Kwara State gubernatorial election an "immoral set of school girls."

From "barbaric" and "savage," the rhetoric moves to "disgusting." The Fom-buen and Ndofoumbgui, ethnic groups in Cameroon, have had their forms of claims making in farmer–grazer conflicts (1980–2010) condemned by some ad-ministrative authorities as inimical to culture, uncivil in disposition, and anti-social. These authorities seek to clothe women ritual participants and get them to adopt "civil" modes of claims making.

The Provincial Delegate for Women's Affairs/Director of Women's Empow-erment Centre Bamenda is quoted as saying, "The Delegation of Women's Af-fairs do not encourage women's associations that operate as secret societies, such as the Takumbeng of Nkwen and the Fombuen as you call it in Babanki. We en-courage development associations. The practices of these secret societies, which operate as associations, are disgusting and expose women's privacy (walking na-ked on the streets)" (quoted in Fonchingong, Yenshu Vubo, and Beseng 2008, 135). In this statement, the provincial delegate shows an intention to neutralize the unsettling effects of insurgent self-exposure. First, the term *secret societies*, inherited from the colonial vocabulary, obscures the important place of simi-lar groupings in the social structure of their society. This bureaucrat hopes to confine "secret societies" to a particular domain of life in the country, making the exposure grounds for repression within postcolonial secular governance, with its promises of democracy and economic development. This confinement is antithetical to the women's understanding of their indigenous religion, which encompasses all spheres of life.

The word *disgust* is reminiscent of the evaluative language that colonial-era missionaries and administrators used to designate most native practices that challenged imported values and practices. And the women are not exposing "their privacy," as the provincial delegate claims, but rather putatively unleashing the maleficent entities that reside within themselves to punish their targets. The question of privacy rejoins that of privatized religion to show the delegate's endorsement, if not the interiorization, of Western-inflected values, of which the claim to secular governance is crucial. The disjunction between the language of genital cursing as religious and the language and structures of the state shows the challenges of deploying the act within the context of the presumed secularized spaces of governance.

Given that no interviews with the women in written or audiovisual form exist (typical of these rituals and forms of protest), it is impossible to unpack their workings against the background of historical, ethnographic, and anthropological research conducted among the women's ethnic groups during the colonial occupation and the decolonizing era. Although illuminating, drawing on these sources is a risky business because their meanings may be at odds with the women's actual thoughts. Reliance on these sources is, additionally, a classic case of mediation, with its possible fault lines.

The public official's statement also obfuscates the important role of these female groups both in anticolonial and postcolonial resistance and in their historiographies. From the 1930s to the early twenty-first century, women's organizations of religious orders from multiple ethnic groups organized protest demonstrations and/or cursing rituals against colonial governance and local despotic rule in Cameroon. Some of these rituals and their ethnicities include the *anlu* of the Kom, the *fombuen* of the Kedjom, and the *ndofoumbgui* of the Aghem.[12] During the colonial era, the protest demonstration, called *anlu* ("to drive away" in the Kom language), stemmed from the women's resistance to the British introduction of cross-contour farming against soil erosion.

Anlu is a disciplinary technique that women mete out to men who commit the following abuses: "beating a pregnant woman or nursing mother, incest, seizing a man's sexual organ during a fight, impregnating a nursing mother within two years of the delivery of the child, or abusing old women" (Ardener 1973, 436–37). The punishment follows a strict sequence: women gather, go to the bush, and reemerge before dawn in war outfits (vines, bits of men's clothing, and faces painted). In the compound of the offender, they dance, sing his offenses, appeal to his ancestors to join the retribution, expose their genitalia, and defecate in the compound, turning it into "a bush." This ritual of ostracism leads to the offender's social or literal death.

On July 4, 1958, the women performed the anlu on several local men at the marketplace, leading to the withdrawal of an unpopular regulation enforced by the British. That regulation consisted in fining dissident women for farming offenses. Then in November of that year, twenty-two thousand half-naked Kom women marched thirty-eight miles to protest the arrest of four of their leaders in relation to anlu activities. However, Emmanuel Konde (1990) argues that indigenous political leaders instrumentalized the women, transforming the anlu from a female-gendered and socially sanctioned disciplinary technique into a political association for factional interests, a rising trend I have explored in section II, "Co-optation."

"Nudity Is Not the Answer . . . What Works Is Legal Papers"

If disparagement is an instance of secularization, so is the call for a "better" form of expressing grievances, one consonant with the demands of political secularism that the postcolonial state aspires to promote. In 2015 Mr. Daudi Migereko, Uganda's minister of lands, argued, "Nudity is not the answer; there are other better ways of resolving women's land issues other than undressing. . . . What works is legal papers" (quoted in Birungi 2015). Yet these papers, the certificates of customary ownership (CCOs), are for a variety of reasons beyond the reach of rural women, as gender and women's studies scholars in Uganda assert (*This Is Africa* 2015).[13] Religious heterodoxy in Africa, however, carries often-conflicting implications for plural legal orders, with defiant nakedness and certificates on two opposed political sides. Perhaps the women consider it futile to get a piece of paper to legitimate what de facto and naturally is theirs. How can a piece of paper bring into being a fact—their landownership, which their ancestors had already been giving them, perhaps for centuries?

However, without access to the women's rationales, goals, and feelings, the need for and quality of translators or mediators are important for the concept of naked agency. Translators constitute mediating figures between women and their local, national, and international targets. Often the mediators' contribution can either enhance or lower women's agency. In response to the minister's comment, Josephine Ahikire, dean of the Department of Women and Gender at Makerere University, had to step in and translate for the supposedly uninformed. She is quoted as saying, "Stripping was a desperate measure because they felt it was the only thing left for them to do and in the end it compelled government to reach an agreement and their land was spared unlike that in Mubende District which today has been sealed off" (*This Is Africa* 2015). Absent from the translation project is the possibility that the minister heard and knew the implications of the women's act. His public statement may reflect his inability to fulfill the women's demands. It is unclear if undressing is what prompted the government to spare the women's land.

On the naked protest by Acholi women in the same wave of land disputes in Uganda, a 2015 opinion piece by Julius Ocungi in the Ugandan *Daily Monitor* prominently features Joyce Bongomin, a bystander-translator-mediator. Described as a fifty-one-year-old resident of Koro Subcounty in Gulu District, Bongomin mediated our access to the meanings of the women's gesture thus: "Elderly women stripping, moreover before children, was to protect what will feed their grandchildren in [the] future, [and] is an act that has been well thought out. . . . Any sensible African would never execute their plan after the sight because women always protect their nakedness" (quoted in Ocungi 2015). With her explanation, Bongomin moves from being a bystander to becoming an active participant in this drama of desperation and rage. These figures are active participants because their mediation-translation affects how (ritual) insurgent self-exposure is to be understood and accounted for. Clearly, the glaring absence of direct elucidation by the women themselves—whose presence in the article is materialized by images of their exposed and blurred breasts and buttocks but not their words—suggests the importance that the bystander-translator-mediator occupies in our account of the protest.

"There Are Several Legitimate and Civilized Ways to Seek Redress"

Alongside or accompanying the rhetorical violence against women protesters are the physical counterattacks, including arrests, threats thereof, beating, desecration, and even murder. In 2002 police and security forces brutalized about three thousand Itsekiri, Ijaw, and Ilaje women of the Niger Delta who occupied Shell and Chevron Texaco facilities in Warri. According to reporters at the time, "The women had a rough deal with security agents in effecting the seizure, leaving ten of them seriously injured. They even beat to a state of coma an Itsekiri woman. The women were teargassed and physically brutalized by the security agents in an attempt to disperse them" (Okpowo and Adebayo 2002). Blessyn Okpowo and Sola Adebayo report that the commanding officer of the Seventh Battalion, Effurun, Col. Gar Dogo, warned that his men were on solid ground to ensure the safety of oil workers and equipment. Additionally, Dogo is reported to have warned the "indigenes" of oil-producing communities against taking the law into their hands, stressing that "there are several legitimate and civilized ways to seek redress" (Okpowo and Adebayo 2002). Perhaps, by experience, the women have learned to distrust "legitimate and civilized" ways to seek redress. Thus, they decided to mobilize modes that may finally get them heard. Eventually, the protest participants vacated Shell facilities after reaching an agreement, the content of which remains undisclosed. The women's abilities to unsettle the quotidian of these multinational oil companies is a victory in itself. In a way, they signal their presence as stakeholders in the exploitation of their communities' natural resources.

The 2016 national controversy after Stella Nyanzi's naked protest against the alleged dictatorial practices of Makerere University officials led several Ugandan government members to call for her arrest for public indecency and pornography (BBC News 2017). Two years prior, hundreds of Togolese women bared their genitalia and breasts to unleash the forces that supposedly reside in their sexual organs. They cursed their dictator president, Faure Eyadema, who had manipulated the constitution to extend his tenure (Abidjantalk 2013). However, Eyadema responded by using his military forces to brutalize and arrest the women for disorderly conduct. Similar murderous responses occurred in 2002 and 2011 in Côte d'Ivoire.

"To Come and See Us Naked, That Is an Action That We Refuse to Tolerate"

In previous instances, it is unclear how the women responded to the disparagement and consequent secularization of ritual cursing or defiant disrobing. It is even unclear how the women framed their act in the first place, given the absence of their words and the frequency with which translators and mediators contribute to making legible their acts. In the 2011 instance of protest in Côte d'Ivoire, the ritual is physically desacralized, which leads to the women's attempts at redress.

In 2011, in collaboration with the Yamoussoukro section of the PDCI, its decades-long ally, the Adjanou women of the city organized a purification ritual in the residence of Houphouët-Boigny, founding father and former president of the country. Curiously, the women's leader is reported as enlisting the assistance of the Christian God (with a capital G) for the success of their ritual: "Yes, our Nation is crumbling and we should pray so that God may save her" (K.A. 2011).[14] A similar combination of references to both indigenous religious practices and the Christian God was manifest in the Gambian women's performance. The goals of the ritual, culled from several newspaper reports, consisted of purifying the country, thereby exorcising the evil spirits (the looming civil war) haunting it. Thus, the ritual participants sought to protect the legacy of the founding father by protesting against the occupation and transformation of his estate into a powder keg by those whom journalists referred to as President Laurent Gbagbo's mercenaries, militiamen, and highwaymen (K. A. 2011; Loukou 2011; Mintho 2011). The purification ritual was mired in controversy and accusations of desecration.

Here, I explore the absence of juridical and constitutional prerogatives that such an organization ought to have received, given its substantive and nominal involvement in the birthing of the African nation-state, as I have demonstrated in my discussion of "Africanizing" women's rituals for nationalist or factional political interests (scene 3).

First, although the dancers had secured the authorization to perform at the former president's estate, when they arrived on location around 5:00 a.m., they were denied access by uniformed men whom the women claimed had "illegally" occupied the premises. After lengthy negotiations, the women were allowed entry, and they started performing the Adjanou. The Adjanou is both the name of the ritual dance and that of the women's society. Securing the permit is a clear example of the intriguing interface of practices of secularism and indigenous religious expressions in the westernized Ivorian context. In that sense, the liberal principle of sociopolitical recognition, in the form of a permit to protest/perform, should protect the women's cultural values and practices. However, in the normalized state of exception and legal ambiguity, several of these liberal secular principles of social and political citizenship may be suspended. Additionally, as the women of the society realized, a permit—the sign of administrative approval—does not necessarily correlate with the juridification of rights or the respectful treatment one is entitled to in particular circumstances.

Second, Mrs. Yvonne Kouamé, the leader of the Movement for the Liberation and Purification of Houphouët-Boigny's Estate and initiator of the week-long dance, scathingly critiqued the Gbagbo regime for desecrating their ritual. Naked, as required by their ritual, the performers were making the rounds at the residence when they discovered to their dismay that a man in a military uniform was videotaping and photographing them. Kouamé said she could not believe that a man, born to a woman, could behave in such a way. The protesters' attempts to confront him failed because a 4 × 4 vehicle with a military license plate picked him up and sped away. Kouamé displayed feelings of desperation and defiance: "We are utterly overwhelmed by this development and we can't take it anymore. . . . Gbagbo has sufficiently humiliated the Baoulé people, and

I believe that this time, here in Yamoussoukro, and with the women, he has reached his breaking point" (quoted in Loukou 2011).[15] Feeling overwhelmed and desperate could have meant the end of the women's goal of participating with their rituals in resolving the postelection standoff; however, they remained defiant, banking on their abilities to mobilize the spirits housed in their bodies. As a response to the desecration of their ritual, the women divided into two groups, alternating between night and day shifts. In the morning, the day shift replaced the night shift after a debriefing on the daily incidents and a program of prayers, incantations, and exorcisms. Recognizing the cohabitation of the supposedly contradictory emotions of desperation and defiance has implications for how we theorize political agency within the suspended juridical system and the frailty of normative structures such as customary values.

In the same vein, the women's uncertainty about the effects of their power led them to reach out to earthly sociopolitical actors as a way to address their grievances. In their quest for recognition of the violation of their belief system, given the blindness of the juridical system despite the issuance of a permit by administrative authorities, the women reportedly implicated various community, political, and state leaders. More specifically, they were described as taking steps to seek reparation for the desecration of their ritual, although the nature of the reparation remains unknown. They swore to strip naked to block access to the estate until they received the explanations and amends they sought: "In the face of this [desecration], we will remain naked in front of the residence until we know why they did what they did.... To come and see us naked, that is an action that we refuse to tolerate" (Loukou 2011).[16] Although Kouamé and the other women threatened their targets with the women's tabooed nakedness, journalists reported that the protesters were fully dressed in white clothes and had their faces smeared with white clay.

Perhaps the nakedness that the organizer sought to expose was less about unleashing the mystical forces housed within their bodies than about the exposure of the violence perpetrated against the women in the form of desecrating their ritual. Recourse to earthly sites of power relations for reparation seems to contradict the belief that otherworldly entities trump all others. To mobilize societal rituals means that ordinary jurisdiction has shown its limits in handling grievances. The postcolonial era, with its constitutional silence on this long-standing indigenous institution of naked protest and violations of women's rituals, has put pressure on and probably exposed the limits of the power effects of women's bodies as articulated by social science scholars and the popular imaginary. The women's need to mobilize invisible forces to redress worldly matters is perhaps already a manifestation of the disintegration of secular state apparatuses. After all, it makes little sense that the women invoke the state, that which they deem incapable of ensuring or unwilling to ensure the rule of law.

The protesters' quest for reparation for the moral injury they sustained through the desecration of their ritual raises a compelling question: how does modern secular governance produce justice for forms of injury intelligible only to specific religious beliefs? A similar question was raised in 2005 during the violent Muslim responses to Danish cartoons representing the prophet Mohammed, which Saba Mahmood (2009) analyzes with an invitation to rethink normative conceptions of religion, law, and language. Mahmood's reflections resonate in the Ivorian context with its myriad religious orders. Different from the seemingly clear divide between the "secular" West and the East, represented by the "religious" Muslims who live in liberal democracies, the Ivorian context pits the aspirational and pompously celebrated secularity of the state against the country's own entangled relationships with mystical forces, and against plural religious (indigenous and foreign) orders. Separation of the secular from the religious, which remains an ideal in our world with its promise of secularism-cum-freedom, takes on a more complex hue in the generalized African context.

Fees Must Fall: Benefiting from Secularized Self-Exposure

Some protesters who marshal insurgent nakedness benefit from the secularization trend because it enables certain kinds of agency. Rather than engage in what would conventionally be called genital cursing, these protesters deploy defiant nakedness by seeking to attract media attention to their cause, shame postcolonial bureaucrats, or voice their grievances. Even without its religious aspect, the secularized version of genital cursing can still be mobilized because it violates the Judeo-Christian precept of decorum. In consequence, naked protest works without the need to invoke the sanctity of motherhood. Such is the case with the naked shaming by female students of the generalized 2015–16 Fees Must Fall movement in South Africa.

Preceding Fees Must Fall is Rhodes Must Fall, which started at the University of Cape Town. In March 2015, students protested for removing the statue of the staunch British imperialist and mining magnate Cecil Rhodes. The call for the removal of the statue is part of the larger goal of decolonizing higher education after the supposed ends of colonialism and apartheid. On April 9, the statue was torn down, galvanizing the student body. In mid-October of the same year, a new set of demands emerged: economic justice for students at the University of the Witwatersrand struggling against the rising cost of higher education. The protest spread to the University of KwaZulu-Natal and to the University of Cape Town, which suspended classes. Students from across the country publicly organized protest demonstrations for more than a year. As their parents did with apartheid, these students consider free higher education to be the most significant issue for their generation. First-generation college students, students in financially strapped households, and international students joined the movement. During the students' protest demonstrations, violent confrontations between them and the police involved the use of tear gas, stones, rubber bullets, and stun grenades (Vilakazi and Swails 2016).

Amid these demonstrations and police brutality, in November 2015, the movement took on sexual and gendered overtones with accusations of rapes and sexual assaults on university campuses, and the alleged dismissive reaction of university officials. Through #RapeAtAzania, protesters expressed outrage at the news of sexual assault of a female member of the Rhodes Must Fall movement; the information was initially posted to Facebook (Mugo 2016). In April 2016, the anonymous release on social media of a list of suspected perpetrators of sexual violence at Rhodes University, known as the #RUReferenceList, sparked protest demonstrations.[17]

The protesters marched with signs (both handwritten and printed) commanding, "Stop the war on women's bodies" or pleading "Listen Hear Us Cry" (figure 6.1). What would later be given the hashtags #EndRapeCulture and #EndPatriarchy demanded an end to practices, attitudes, and behavior that normalize sexual violence on campuses. Prosecution of alleged rapists, who reportedly were going free, is a step toward meeting that demand. The movement was responding to local and national trends: the widespread sense of insecurity for women, unconventional genders, and sexual behaviors. In 2015 South Africa recorded more than 43,195 incidents of rape—the highest in the world.

A few days later, hundreds of black and white female protesters made history by baring their breasts, thus initiating the trend of cross-ethnic naked shaming within the movement. Their shaming strategy inspired other students, like those of the University of the Witwatersrand.

In October 2016, in their efforts to resist by defeating repressive police tactics, three female students went topless and held their hands on their heads. In one SABC news broadcast posted to YouTube, the women are heard pleading, "Cease fire, stop beating us, stop beating us," "We are your children, stop shooting us," "We are not armed, stop shooting at us." One bare-breasted female is distinctively heard pleading, "You beat me up this morning and my hand is swollen, stop, stop" and "I'm just a student coming to campus and you beat me up" (SABC Digital News 2016). It is unclear whether the students stripped on

FIGURE 6.1 Rhodes University students protesting rape culture in April 2016.
Courtesy of *Sunday Times* (Johannesburg).

the scene or came to the scene already topless. They are facing the police, who are almost off-camera; the focus is on the women. Around the protesters is a large crowd, most of whom are busy videotaping or taking photographs. Some of those present hold professional-looking cameras, suggesting that they are journalists or news reporters. It is also unclear whether the crowd is composed predominantly of witnesses, bystanders, or protesters. With the viral nature of recording, witnesses, supporters, and sympathizers multiply indefinitely. To consider social media as radically positive in the performance of uncivil disrobing is to overlook its possible pitfalls. For instance, the speed and ease at which images circulate and the frenzied archiving tendency in our Era of Life Behind Screens indicate that these women may in future years regret revealing their tabooed body parts in public (as I discussed in scene 2 through the documentary *Uku Hamba 'Ze (To Walk Naked)* [1995]).

In the YouTube video, the baring of breasts registers no physical violation by the police. The main outstanding form of injury is that suffered by the women who had to resort to self-exposure to protect their bodies: violence to the self in order to protect the self from possible death. Ignoring the possible later implications of their act, especially in the presence of innumerable recording devices, the students offer their bodily vulnerability and their pleas as the platforms over which struggles for authority and power are to be waged. Their corporeal vulnerability expresses their defeat in the face of repressive violence.

In relation to the naked demonstrations of the Rhodes students against rape culture, these Witwatersrand protesters supposedly reversed the power dynamic, the need to subdue the opponent, that often undergirds instances of rape. By assuming authority over their bodies, by deciding when to go covered or when to disrobe, these women undermine the potential rapist's decision-making power. Paradoxically however, their toplessness may signify vulnerability, which will curtail their targets' sense of accomplishment. Given that rapists are claiming for themselves the right to dominate others' bodies, what does it mean to dominate that which is already available to dominate? Clearly, the women in these instances are not mobilizing the spiritual and/or religious aspects of disrobing in anger. Perhaps they are mobilizing the enabling effects of the disseminating frenzy of our media-saturated world.

Although the secularizing project benefits certain protesters, it also markedly undercuts the implication of the women's gesture. This dual process of benefit and curtailment becomes manifest in ritual cursings, called Opération Kodjo Rouge, in Paris, that encounter the silence of the French political class in a form of epistemic ignorance.

Epistemic Ignorance and
Menstrual Rags in Paris

"White ignorance" was meant to denote an ignorance among whites—an absence of belief, a false belief, a set of false beliefs, a pervasively deforming outlook—that was not contingent but causally linked to their whiteness. —**Charles Mills**, "Global White Ignorance"

In July 2011, at the Trocadéro Human Rights Plaza, a popular tourist site in Paris, hundreds of Ivorian women brandished red cloths to curse American, French, and Ivorian political leaders for the intervention of the French/United Nations army in the 2011 postelection conflicts in Côte d'Ivoire. The cloths supposedly were menstrual cloths that the women, for symbolic impact, insisted they had worn. The French/UN intervention led to the arrest of former Ivorian president Laurent Gbagbo, who was held for years at the International Criminal Court (ICC) in the Netherlands for crimes against humanity. In January 2019, Gbagbo was released after his acquittal. For five years, this women's collective, called Opération Kodjo Rouge, moved beyond the boundaries of the postcolonial state to organize events in Paris for purposes of decolonization and political autonomy. In addition to cursing (invoking the power of menstruation, exposing their breasts, taking the Earth as witness, and dancing to awaken ancestral spirits), the participants actively mobilized social media to make claims. But it was their hope to produce both figurative and literal deaths of then-presidents of the United States and France—Barack Obama and Nicolas Sarkozy, respectively— that met with the silence of "the French political class" and is worth analyzing for the insights it provides about the texture of women's agency.[1]

Drawing on social media material and newspaper reports, this scene evinces a particular kind of political subjectivity that we see emerging in postcolonial biopolitical Africa. The ritual cursings powerfully contribute to the idea of naked agency. I argue that agency is not located in a specific site but rather lives precariously in a praxis of perpetual movement and resists unitary and fixed paradigms; this is so because the women's presentation of their rallies as customary and African reinforces the interpretation of their cursing as exotic and primitive. I highlight expressions of paradoxical agency that lie in plain sight and yet are not comprehended with less than capacious reading lenses, a point akin to Gayatri Chakravorty Spivak's argument in "Can the Subaltern Speak?" (1988). The women's cursing rallies are similar in goals and yet different in methods to sociopolitical acts that include physically violent and terrorist strategies ranging from throwing food at politicians to inflicting lethal indiscriminate violence on civilians or elected officials. Yet, while the latter acts are understandably taken seriously, the former are seen as nonthreatening and therefore ignored or dismissed in a paradoxical dynamic inherent to globalization.

Globalization, the effect of which is the increasingly connected world, has heightened the visibility of specific forms of contestation. Yet, given the standardization of cultural, political, and social norms, globalization also represses modes of political dissent not consonant with the Enlightenment-inflected and bourgeois-informed channels of participatory democracy (including voting, participating in political parties, lobbying, and writing letters to politicians). Although visible, the women's cursing rituals in Paris, which they markedly modified to respect legal limits of nakedness, encounter the epistemic ignorance bequeathed by structural privileges to French authorities and press who remain silent on the rallies. By appealing to their "traditional" punitive strategies, participants in the rallies also intensify the long-standing view of Africans as primitive and unable to adopt modern forms of being and acting, as Nicolas Sarkozy suggested in 2007. Nevertheless, silence about the most violent acts performed by the women (from the women's perspective) suggests how globalization both reads and represses its own colonial histories.

Such a silent reaction was perceivable in Sarkozy's infamous Dakar speech of July 2007, of which parts share similarities with the Hegelian conceptualization of Africans that I have analyzed earlier. In the speech, reproduced in *Le Monde*, Sarkozy argues: "The tragedy of Africa is that the African has not fully entered into history. . . . They have never really launched themselves into the future. . . . The African *peasant* only knew the eternal renewal of time, marked by the endless repetition of the same gestures and the same words. . . . In this realm of fancy there is neither room for human endeavor nor the idea of progress" (2007, emphasis mine). Why is it the *peasant* that Sarkozy conjures up for his conceptualization of Africans? What should we make of the adverb *really* in "really launched themselves into the future"? And, most importantly, how do Africans who live in Europe fit into Sarkozy's conceptualization? Even today, African politicians and intellectuals continue to reflect on or to seethe over these statements.

The former president's speech ignores the active ways in which Ivorian women, in particular, from the colonial era to the neo/postcolonial moment, have challenged Sarkozy's account of Africans as having never really launched themselves into the future. Earlier, I highlighted the opposition against French rule that Ivorian women mounted during the struggles for independence. In the neo/postcolonial era, they continued to challenge overtly French political and administrative leaders, starting in 2003 with Dominique de Villepin, France's minister of foreign affairs.

Abidjan 2003: Dominique de Villepin and Exorcism

I trace the public and political deployment of Opération Kodjo Rouge in Paris back to the January 2003 incident in Abidjan when a group of Ivorian women known as the Amazons of Auntie Simone, who were under the wing of then–First Lady Simone Gbagbo, challenged French Minister of Foreign Affairs Dominique de Villepin.

The larger political context that informs the incident is the multipartite power struggle involving France, Ivorian rebels, the Ivorian opposition, and the ruling party of Laurent Gbagbo over the French-sponsored Linas-Marcoussis Agreement (January 15–23, 2003).[2] In 2001 the regime of Laurent Gbagbo had come under attack by rebels led by Guillaume Soro Kigbafori, resulting in a partitioning of the country into three zones: the north, under the control of rebel forces; the south, controlled by governmental forces; and in between "the zone of confidence," held by the French army. Although French companies allegedly sponsored the rebellion to destabilize the anti-French regime of Gbagbo (Diop 2005; Frindéthié 2010; Tayoro 2002; Union Comunista Internazionalista 2003), French president Jacques Chirac and de Villepin invited the Ivorian political class to meet in Linas-Marcoussis (France) under the aegis of the French Cabinet of Foreign Affairs. The meeting's purpose was to draft a peace agreement and a plan of reunification of the country. After complex and prolonged negotiations, Gbagbo and his party agreed to open the government to the opposition, the rebels, and to the leaders of the so-called civil society. However, upon returning to Côte d'Ivoire, Gbagbo's supporters organized massive protest demonstrations to oppose the Linas-Marcoussis Agreement and what they perceived as France's neocolonialist maneuvers. These demonstrations, reportedly masterminded by political figure Mamadou Koulibaly and First Lady Simone Ehivet Gbagbo, fed Gbagbo's refusal to implement the agreement, thereby stalling the peace process.[3]

On January 3, 2003, de Villepin visited Côte d'Ivoire to reactivate the agreement, meeting with Gbagbo for half an hour. However, the minister was prevented from leaving the presidential palace by protest demonstrators screaming, "On veut Gbagbo" (We support Gbagbo) (de Chalvron 2003). Some even called the French minister a terrorist: "Villepin assaillant, terroriste, on veut Gbagbo" (Villepin, aggressor and terrorist, we support Gbagbo) (quoted in Duval 2003). For forty minutes, de Villepin was detained; footage produced by the French television station France 2 shows him brimming with impatience and frustration.

The footage, however, makes no mention of the major reason for de Villepin's frustration: his confrontation with naked members of a women's society who urinated on his car to "exorcise" the devil, the French, from Côte d'Ivoire (*La Lettre du Continent* 2003).[4] While France 2 "unofficially" censored that component, newspapers and magazines brought it to the attention of the general public. Reportedly, two dozen Gbagbo female supporters had stripped naked, their bodies covered with white clay, and urinated on the car that was driving the French minister. Observers suggest that the First Lady—fondly called Auntie Simone and someone allegedly adept at all sorts of indigenous practices—masterminded the ritual performance (*La Lettre du Continent* 2003; Malagardis 2012).

Such pseudo-indigenous practices are popularly described by terms such as *fetishism*, *maraboutage*, and *prophecy*. The imbrication of fetishism in politics is a long-standing documented practice on the continent; although it remained underground following the independence years, it later became ubiquitous in the postcolonial period (Bernault 2005; Buijtenhuijs 1995; Frère 2000; Geschiere 1995; Janin and Marie 2003; Martin and Laplantine 1994; Naipaul 1984; Séhoué 1997; Tonda 2005). These strategies hold the potential to backfire because the women's acts may be perceived as prepolitical and primitive. Questions then become: How does one resist in different epistemic orders? What happens to the resisters' agency if their targets misinterpret their gestures? It is in light of such misunderstanding, either intentional or otherwise, that naked protest is inherently precarious.

Eventually, President Gbagbo joined de Villepin outside the presidential palace to pacify the hostile crowd. According to journalists, the presence of ritual exorcists or protesters, depending on one's vantage point, instilled fear in everybody, including security forces. The widely respected African magazine *La Lettre du Continent* (2003) explains the implication of the incident thus: "All African presidents fear this form of protest, and even their security guards are of little help, because they cannot look at the naked women who symbolize their mothers and who urinate to chase evil spirits."[5] This explanation of the ritual is consonant with the speculation of the Ivorian Women of France's blogger Louis-Freddy Aguisso (2012), who asserts that the misfortunes—fall from power and illness—that befell de Villepin following this very tumultuous Ivorian visit were due to his being cursed by the exorcists.[6] As expected, journalists and bystanders provided the meanings of the ritual because the sacred nature of these acts prevented the actors themselves from interacting with the general public. There is more to the use of the term *sacred* to frame this performance. In addition to the sanctity of the curse is the political mileage to incite fear in opponents by packaging, read "Africanizing," the protest as a sacred performance. With that framing comes the normative expectation that journalists and "secular" spokespersons mediate the implications of these rituals.

This incident with de Villepin, which might have been condemned in other contexts as insulting, is, however, Africanized, Ivorianized, or "nationalized" by the Gbagbo camp, first by the overdetermined rhetoric of motherhood and then by "tradition," a point I made earlier. For instance, the women's ritual performance was explained as the Gbagbo regime's investment in encouraging women's celebrated participation in national political processes. That narrative is akin to the United Nations' developmentalist program "Gender and Development"—critiqued by Spivak (1996) and others—whose rhetoric is then converted into political capital in the international arena.

The clash between de Villepin and the women protesters is not an isolated incident. In the colonial era, similar encounters between colonial administrators and the natives were numerous. These incidents remind us, if need be, that the colonial and the postcolonial share as many similarities as differences. Further, these encounters involving genital shaming or cursing may have contributed to colonial administrators' accounts of women as uncivilized, thus weakening the women's agency. With de Villepin's confrontation, two epistemic orders are forced into an almost impossible dialogue: his diplomatic visit with its putatively secular aspects, and the mystical message, written on the female body with nakedness and exorcism as its pen and signifier. Whereas the foreign minister's message seeks peace via diplomacy, the ritual exorcists' action aims at peace via injury. In fact, exorcism by the initiated—the guardians of the land—and the diplomatic visit both stand as two modes of conflict management, differing in strategies yet converging in goals. This rowdy meeting materializes the generative encounters between two worlds, not fundamentally opposed but rather framed as mutually exclusive.

Opération Kodjo Rouge: 2011–2014

Since 2011, the name *Opération Kodjo Rouge* has become part of the Ivorian political landscape because of an organization of Ivorian women in France.[7] Opération Kodjo Rouge is a relatively new form of social punishment and claims making. Its emergence in France to support the former Ivorian Bété president Laurent Gbagbo demonstrates the growing ethno-politicization of gendered modes of action in and about Côte d'Ivoire. Unlike Opération Kodjo Rouge, the Adjanou, the Akan female society, has been prominent on the political landscape since the pre-independence period of the 1950s. That ethno-politicization, already perceptible in the footnotes of historical accounts, continues to shatter the elite nationalist fantasy of homogeneity during de-colonization and in the postcolonial era. These women-centered actions have undergone massive restructuring and modifications to reflect the interests of postcolonial bureaucrats, and the movements are "nationalized" by different regimes and parties, as articulated more centrally earlier.

Eight years after the 2003 diplomatic incident involving exorcism through cursing, the practice resurfaced, this time in France but again in support of Laurent Gbagbo, whom the ICC in The Hague was charging for crimes against humanity. That Gbagbo's supporters would demonstrate against his indictment came as no surprise; however, the shocking aspect, especially in France, was the deployment of their menstrual cloths.

In July 2011, hundreds of women of mainly middle and old age, all dressed in white outfits (T-shirts, shorts, leggings, trousers, or jeans), met for a unique form of rally (Togui 2011a). The women's faces and upper bodies were conspicuously made up with white clay in various designs. They wore striking red loincloths, "kodjos" in Baoulé, over their pants (figure 7.1). White and red are the dominant colors of the ritual; red refers to menstruation, and white refers to peace and/or death. The participants sang, chanted, screamed, and hit the pavement to the rhythm of the chant, "Alassane ooh, Sarkozy ohhh" (the "ohhh" communicating their pain). On a podium, some used loudspeakers to galvanize the women in their acrobatics and wailings, leading to a frenzied atmosphere. Several dancers exposed parts of their breasts in syncopation with the chanting, which event bloggers later labeled as "cursing."

MAP 7.1 Ethnic and political map of female collective action, Côte d'Ivoire. Courtesy of Tim Stallmann.

FIGURE 7.1 Opération Kodjo Rouge I, July 24, 2011. Source: Civox.net.

Topo Léontine, the women's leader, organized Opération Kodjo Rouge rallies. According to Topo, the women used Opération Kodjo Rouge to curse designated targets on the Ivorian political scene (Togui 2012c), including Alassane Ouattara (president of Côte d'Ivoire) and Sarkozy (then-president of France), whom the women identified as criminals, assassins, and rapists—in sum, those who have plagued the country since the beginning of the armed rebellion in 2002. The first Opération Kodjo Rouge inspired four more in Paris:

Kodjo Rouge II (September 2011),
Kodjo Rouge III (March 2012),
Kodjo Rouge Terminator (October 2012), and
Kodjo Rouge International du Tonnerre (June 2014).

Initially, the cursing rallies were to form a trilogy; however, after the first three—galvanized by the defeats of Abdoulaye Wade and Sarkozy in the Senegalese and French national elections, respectively—the coalition organized the fourth and supposedly deadliest one, Opération Kodjo Rouge Terminator. "Terminator" unmistakably borrows from the massively, internationally popular science fiction film franchise of the same name, thereby participating in the global cultural flows of the term. Unfortunately, this fourth rally did not live up to its namesake's popularity; a fifth rally, International du Tonnerre

FIGURE 7.2 Opéra-tion Kodjo Rouge III, March 2012. Women cursing Sarkozy's and Ouattara's pictures with their menstrual rags. Source: Civox.net.

(Thunder), became necessary given the unimpressive effects of Kodjo Rouge Terminator (AFP 2012).

According to organizers, Kodjo Rouge II at the Place de l'Opera in Paris drew approximately one thousand male and female participants. It was supported by leaders of multiple political and social organizations, according to one of the organization's main bloggers, Zéka Togui (2011b). Its goal differed slightly: this time, the goal was to free Laurent Gbagbo and other political prisoners in both Côte d'Ivoire and France. Participants denounced what they call the military coup of April 11, 2011, in Côte d'Ivoire; the arrests and imprisonment without trial, the torture, and the kidnappings of women and men; and other human rights violations, including the denial of fundamental freedoms to civilian populations (Togui 2011b). The atmosphere was cruder in terms of women's songs, incantations, chants, and acrobatics.

Kodjo Rouge III was more festive and diverse in terms of nationality, gender, and modes of entertainment (figure 7.2). In the third rally, the participants chanted "Sarkozy assassin" while hitting the pictures of presidents Ouattara and Sarkozy with their stained menstrual cloths. The French police reportedly attempted, unsuccessfully, to prevent ritual participants from chanting those words (Togui 2012b). Bloggers for the movement described several women as being in a trancelike state, weeping and distraught, which attracted curious onlookers. Blogger Togui wrote that tourists encountered, and took advantage of, an "exotic spectacle" (2012b). The association of terms such as *exotic* and *spectacle* suggests the limitations of this temporary relocation and reconceptualization of *kodjo rouge* in Paris. As the collective continued to perform their rituals and engage social media and other reliable mechanisms of social movements, they succeeded in creating a public spectacle that drew media attention from professional journalists—but usually only Ivorian journalists.

While the first two Kodjo Rouge rallies were covered and publicized primarily in social media, Opérations Kodjo Rouge III and Terminator attracted the attention of Ivorian journalists, including Jean Paul Oro and Lacina Ouattara. As suggested earlier, journalist Ouattara (2012) argues that participants in the Opération "Kodjo Rouge" in Paris made fools of themselves. Unlike Ouattara's scathing op-ed on the cursing, a supportive piece by Jean-Paul Oro (2012) frames the women as patriots resisting the unfortunate encroachment of neocolonial powers.

In 2014, in an interview uploaded to YouTube and transcribed on her organization's blog, event leader Topo candidly reflected on the poor effects of previous rallies, acknowledging the silence and the refusal of their targets to satisfy their demands (Togui 2014). Independently of the effects of their cursing, Opérations Kodjo Rouge invite a rethinking of the question of agency in social movements that utilize direct-action strategies such as sit-ins, strikes, and blockages. What kind of agency is at play in acts involving the actors' menstrual cloths? Further, how effective is a protest demonstration when the targets misread it?

Agency and the Weapon of the Desperate Powerful

The name of the cursing rallies, Opérations Kodjo Rouge, is generative. *Kodjo rouge* is an expression that comes from Agni and French words: *kodjo* (Agni, an Akan language of Côte d'Ivoire) and *rouge* ("red" in French and "stained" in this context). Ivorian literary critic and linguist Germain Kouassi translates *kodjo* in French as "underwear with no link between the thighs and the buttocks, a loincloth" (2007, 76).[8] However, Kouassi's translation is a limited and sanitized version (no mention of menstruation) that fails to take into account the cultural associations of the kodjo in contemporary Ivorian parlance.

Although the term *kodjo* has been systematically absorbed into Ivorian French, its original association with menstruation, primitivism, fetishism, backwardness, and occultism persists, shaped by generational and class configurations.[9] Although older people show respect for and fear of the kodjo, they do not necessarily associate it with primitivism and backwardness but rather with the female principle and power.[10] Yet among young people, occultist resonances of the kodjo supposedly reflect the backwardness of women of rural areas and lower social economic classes who supposedly continue to wear the kodjo, and their failure and/or inability to adopt the obvious signs of modernity by wearing panties. The confinement of the clothing item kodjo to tradition, whereby it is antithetically positioned to its equivalent modern panties, was already intro-

duced by one of the organization's bloggers, Zéka Togui. Reporting on Opération Kodjo Rouge I, Togui briefly traces the origin of the kodjo, claiming that
it is the ancestor of the G-string: "The kodjo finds its origin in the pure African tradition. It is an overwhelmingly red piece of cloth used by a female as a
G-String [posing pouch], before Western modernity with its panties enters in
the mores of African women. As a loincloth, the kodjo enters in the symbolic of
feminine modesty, even if it exceeds that. It also symbolizes the female power
to curse, the female who gives life, and whose word is sacred when she takes off
and brandishes the kodjo to curse" (2011a).[11] Whereas the organizers and their
bloggers draw on the power of the kodjo, others regard the garment as indicative
of problems of transition to modernity and its conventional underwear. If such
is the case of the kodjo in Côte d'Ivoire, Opérations Kodjo Rouge in Paris may
undermine the women's agency if their cursing is read as backward.

   Given that older women's genitalia and their menstruating capacities, or lack
thereof, provoke fear and repulsion among Ivorians, the kodjo rouge becomes a
powerful entity that often carries within it signs of abjection. The kodjo's perceived ability to harm explains these emotional reactions, which suggest that
the reactions are based on the interpretation of the meanings of menstruation,
shaped overwhelmingly by larger societal interpretations.

   The interconnectedness of cultural practices on the continent explains how
Togui's explanation resonates with the ongoing potency of Yoruba women's
insurgent use of their menstrual cloths, which the Canadian psychiatrist Raymond H. Prince analyzed almost six decades earlier, in 1961.

> The menstruating woman and the witch both have power to render
> magic and the native doctor's medicines powerless. . . . [The women's]
> power was so great that they forced the ruling chief of the town to retire
> to the provinces for a year. They camped in hordes in front of the chief's
> palace singing and causing a disturbance. When police were sent to
> disperse them, the women brandished their menstruation cloths. This
> caused the police to take to their heels, for it is believed that if a man is
> struck by a woman's menstrual cloth he will have bad fortune for the rest
> of his days. (798)

Thus, the menstrual cloth, here kodjo, becomes an ontologically animated object with its own power, the very definition of fetishism. The word has come to
stand for the thing (an agent itself), its effect, and the agent (the person) that
wears the agent. And this thing-agent possesses a biography of its own, an identity that may be lost as it travels from the rural settings of Côte d'Ivoire to the
pavements of the Human Rights Plaza in Paris, the so-called capital of political

rationality. However, the organizers and their bloggers attempt at all costs to keep the biography alive (via tradition, menstruation, and cursing) without distorting it, as the thing-agent travels both geographically and temporally, which becomes crucial to legitimizing their "Ivorianness," as putatively imagined Ivorian ways of being in the world are mobilized by women for an Ivorian issue. The Ivorianness strengthens the women's mission of assaulting French values, thus highlighting the failure of the civilizing mission.

Two widespread scholarly accounts of social movements—both shaped by Eurocentrism—have limited our understanding of the wide range of profiles and positions of those involved in contestation and resistance. The first account, disseminated through James C. Scott's seminal study *Weapons of the Weak: Everyday Forms of Peasant Resistance* (1985), frames contestation and resistance as the exclusive prerogative of the powerless, the weak, the injured, and the oppressed: in sum, those whose voices should be heard solely in situations of fighting back. This account overlooks the power that the women here feel and express, as well as the fluctuating dynamic of forces inherent in their situation. The second account understands social movements as bereft of gods and spirits, in other words, as contexts where the idea of a curse will not be thought of in conjunction with political agency or with social movements writ large.[12] I refute these two accounts by showing the kind of agency at play in Opérations Kodjo Rouge.

Contrary to the long-held assumption that those engaged in collective contestation are always already the weak and powerless, the participants of Opérations Kodjo Rouge positioned themselves not as mere protesters (those putatively in the weaker position) but rather as righters of wrongs, as those taking seriously their responsibility as custodians of the community. In radio and newspaper interviews, Topo for the organization explained the workings, meanings, and potentialities of her strategy for her group's explicitly political agenda. In the social structuration schemas[13] of the Bété of Côte d'Ivoire that Topo holds valuable, women are considered guardians of life, the stronger entity in a man-and-woman dyad in the symbolic realm, and those with the capacities to alter challenging communal circumstances. In sum, these ritual participants possess "a tremendous power to curse those responsible for a crime or an injustice, to put them out of action for good" (Togui 2012c). After the March 2012 ritual Opération Kodjo Rouge III/International, Topo is quoted as saying, "This is the opportunity for us to act for the defeat of Nicolas Sarkozy. The curse of the Kodjo will befall him and he will lose the upcoming presidential elections" (quoted in Oro 2012).[14] The curse will materialize thanks to the authorizing cultural belief in the nature of menses and, by extension, of women's genitalia, which hold potent mystical energies with unparalleled abilities to give life. However, the forces (maleficent or beneficial depending on the circumstance) lie dormant in the female body until they are activated through childbearing, initiation, and maturity.

Using generalizing categories of "African," "traditional," and "woman," Topo continues to elaborate on the specificities of the kodjo rouge curse and sets up a stark opposition between them and the French political class. These events may be organized individually or collectively, and the targeted entities may be members of the same uterine household or public figures. In fact, cursing requires the following six conditions for its impact:

1   women, most specifically mothers, as the active agents;
2   the public exposure of stained menstrual cloths;
3   desperate circumstances;
4   incantations that designate the target(s);[15]
5   invoking the Earth that houses all; and
6   that the targets be born to women.

During their cursing rallies, the participants used their "stained" loincloths to hit the ground by uttering curses upon their targets. By hitting the ground, these desperate yet powerful women mobilized the destructive forces of the universe by taking it as witness. This mobilization of the creative forces of womanhood, of words, of ancestors and spirits, and of the earth undeniably orients forces that would miss their targets only if those are not born to women: "If only you are not born to a woman, you would escape the wrath of a woman in this situation."[16]

The position of the women is generative as well as intriguing given the multiplicity of perspectives that ought to interpret this drama of authority and desperation. From an outsider's perspective, these cursing rallies show the women's powerlessness. However, that unitary script fails to account for the complexity of the situation, because from the women's own perspective, their position of authority as guardians of the community mitigates their feeling of desperation. Louis-Freddy Aguisso, one of the Opérations Kodjo Rouge bloggers, specifies that recourse to cursing with the kodjo rouge reveals women's lack of other options, making it an act of both defiance and desperation: "They (women) counterattack with a means that works as their lifeline, as the custodians of life and given their position as the mothers of men" (2012).[17] Thus, those in positions of power or those who violate the law of the land are not immune to the forces of the universe when so invoked. In this case, these actors do not fundamentally conform to the conventional script of the weak and the powerless, and their cursing rallies weaken the dominant narrative that constitutes embodied forms of contestation as the weapon of the weak and/or powerless.

If the cursing follows the customary trajectory, it implicates witnesses in powerful ways. For example, among the Yoruba, the role of witnessing carries deep and abiding responsibilities of sharing the same space at the same time with the women (Oyeniyi 2015). Taking the world as witness works to enforce the efficacy of the cursing by seeking to ostracize socially the women's targets, which may result in their literal deaths although no such deaths have been substantiated. If enlisting the contribution of others for a cause suggests one's powerlessness, it is also a sign of moral force or of cultural capital. The capital is to be in a position to mobilize others for questions that may or may not be of direct interest to them. Given their ability to summon others, the women show themselves as powerful entities, although their authority is problematized by an anatomical determinism.

Conventional studies of resistance and collective action have seldom paid attention to the intervention of gods and spirits in social movements. Thus, widespread assumptions posit that legitimate, respectable, and perhaps "studyable" social movements are bereft of spirits and gods. In analyzing forms of contestation that draw on gods and ancestors, my study joins the conversation started in postcolonial studies by Ranajit Guha and Dipesh Chakrabarty and their exploration of these questions in the South Asian case.[18] During Opérations Kodjo Rouge, the women were desperate yet powerful because of their abilities to mobilize multiple entities (menses, Earth, and divinity) to punish their targets. In that sense, a binaristic script of the powerful versus the powerless fails to capture the complexity of the situation. Which accounts of agency best describe the intervention of these women in the political restructuring of their country? Are these ritual participants only channels through which wrongs are righted, and thus devoid of human agency?

Topo Léontine foresees such a reading and adopts a preemptive move by refusing to reduce women's importance to the spiritual realms of ancestors (Aguisso 2012). To do so, she highlights the necessary courage and the social activities that allow a woman to be qualified to enlist the assistance of spirits. Even if women's genitalia are construed as housing powerful entities, she explains, the woman still needs to activate and maximize these entities through initiation and childbearing. To compound Topo's point, Aguisso (2012) points to the psychological strengths (mind) necessary in this scenario. Reporting on Opération Kodjo Rouge International, he explains: "That's why, when a woman faces a blatant and irresistible injustice, she resorts to that which in her body, but also in her *mind*, is her strength. And, her *strengths* lie exclusively in her *sex*. So in

Africa, a woman publicly exhibits her nudity only in extreme circumstances—such as death or to utter a curse on someone or something—but never for publicity" (Aguisso 2012, emphasis mine).[19] At this point, it is needless to point out Aguisso's generalizing rhetoric in referring to the continent as a whole and to women independently of their age, maternal status, or educational and professional levels. Most importantly, Aguisso's attempt to inscribe the mind in the deployment of the kodjo rouge is confusing, as the mind is reduced to sex: "And, her *strengths* lie exclusively in her *sex*." Independently of these confusions, from the perspective of the primary organizers of Opérations Kodjo Rouge, the mystical powers remain insufficient without women's determination to mobilize them. Further, the necessary awareness of their dire circumstances and their understanding of the situations that demand their intervention—combined with their courage to curse and to enact change—deserve recognition more than the presumed mystical powers of their bodies.

In addition to allowing the return in Paris of the repressed expression of spirits, the women's naming of their rituals paradoxically shows the failure and the success of the civilizing mission. With their term *Opération*, a decidedly militaristic term, they prove, contra Audre Lorde's (1983) argument, that the master's tools can slowly dismantle the master's house. Paradoxically, however, the name, in a highly Calibanesque move, also suggests the success of the colonizing enterprise. Similar to Caliban, the archetypical colonized figure in Shakespeare's *The Tempest*, who cries out, "You taught me language; and my profit on't / Is, I know how to curse" (1908, 19), the women have been taught the language of "operation" through the multiple violent operations of the French army in Francophone Africa. The last French operation that the organizers denounced is Force Licorne's Operation Unicorn of 2002–3, during which the French Armed Forces supported the United Nations Operation in Côte d'Ivoire (UNOCI) to unseat President Gbagbo. As excellent students of their "masters," the women have learned how to curse, not only in their own language and cultural context but also in the language of their oppressors. As such, Opérations Kodjo Rouge remind the French that cursing is a weapon *and* an operation.

In exploring the implications of the term *opération* and its attendants *kodjo* and *rouge/red/stained*, the women engaged in the weaponization of names, reprising Mary Louise Pratt's (2009) concept "weaponization of psychology," which she explored in the context of the U.S. military intervention in the Middle East.[20] The women's employment of the term *opération* is useful because it suggests that the mightiest army of our time does not possess the prerogative of the power of language. More importantly, historically marginalized actors can bring to our attention other ways of weaponizing language and by extension psychology. With their name, the organizers of Opérations Kodjo Rouge suggest their organizational skills, goal setting, and determination, all being characteristic of military operations, without disavowing their beliefs in the potency of words to create reality through incantations.

Inciting and intimidating the opponents are strategies as old as war itself. As a name, "Opération Kodjo Rouge" fits into the weaponization of the power of language; the choice of words is not just reflective of the war-making state but is also constitutive of it. The name not only recoups the term from their opponents but also sheds more light on these women's remarkable creativity because the name is coded with an eye toward public relations, which speaks to their goals of acting in multiple registers: indigeneity with the call to ancestors; and global audience with a bilingual naming, "operation" and "kodjo." Further, the name materializes the coming together of deceptively conflicting worldviews and frameworks: French and Agni, rational and irrational, mind and menstruation, operation and female genitalia. Rather than sanitize menstruation of its conventional trappings of hormones (and, thus, perceived irrationality) and natural history, the ritual participants attempt to activate the threatening attributes of menstruation with the concept of "operation," with its heavy military and organizational connotations.

Coupling the term *opération*—loaded with its militaristic and technological connotations—with the terms *kodjo* and *rouge* belies any attempts at peacemaking, at least by nonviolent means. The conventional association of the color red with blood and violence is not a new phenomenon. Additionally, the relation of red to menstruation in this case is unmistakable. Although menstruation in the Western context has been framed in polluting terms, that understanding has been immensely decentralized by feminist and womanist scholars who regard it as a necessary biological process and not as a source of shame for women. Womanists and cultural feminists have underscored the dialectical nature of menstruation as both life giving and life taking (Madhavan and Diarra 2001; Washington 2005).

Silence, Silence, and More Silence

Despite supposed biological sharedness that undergirds the faulty argument that the exposed body carries the same impact across geographical and temporal spaces, the silence of the women's targets became deafening. In May 2014, Topo Léontine announced Opération Kodjo Rouge International du Tonnerre and spoke candidly about the silence of the French political class and press: "We are realizing that people don't want to listen to us and they continue to keep people in jail. They still have our president Laurent Gbagbo in prison. Blé Goudé just joined him at the ICC. And Mrs. Simone Gbagbo is still in jail" (Togui 2014).[21] The stalemate in their country contributes to the ritual participants' dire need for political recognition. However, what if the stalemate results from the silence of their targets, silence that itself represents the misunderstanding or the misreading of the effects of the kodjo rouge? How can an act that calls for the deaths of two of the world's most powerful leaders go unnoticed if the cursing was taken seriously? In the absence of any public pronouncements from the women's targets, I argue that the silence is a form of epistemic ignorance.

Although literary fiction cannot be confused with historical reality, a brief reading of a novel reimagining the colonial contact shows how the silence of the French political class and media demonstrates a pattern of privilege and asymmetric ignorance that contributes to reifying structural hierarchies. In the Ivorian novelist Jean-Marie Adiaffi's *The Identity Card* (1983), to pick a single example from a long series of possible reimaginings, the third-person omniscient narrator reports an affectively charged conversation between a colonial official and his African aide on a women's ritual purification dance. The conversation reveals the mental state and reaction of the French colonial commandant: "Commandant Kakatika himself was rather worried in the face of the strange happenings taking place in front of his somewhat stupefied eyes. He kept on asking his African companion questions: 'What are they saying? What the hell are they saying in that damn primitive dialect of theirs?'" (9) to which the African responds, "terrible and terrifying things." The commandant insists: "But what on earth are they doing? What are they talking about? What the hell is happening over there?" "Sir, some really strange things are happening" (9). "'What the hell are these dances of savages about?' insisted Commandant Kakatika, who has neither the time nor the mind to appreciate the beauty of all these naked women under the blazing sun" (10). Despite his insistence on knowing "what the hell is happening," Commandant Kakatika dismisses the explanation of his aide.

This description of a white man with military power witnessing opaque behaviors by women in Africa shares similarities with contemporary white men in and with power in Paris. The latter discipline "African women doing strange things" through the presence of the police but remain overwhelmingly silent. There are possibly two reasons behind the silence: the opacity of the cursing rallies and ignorance of ways to respond appropriately to menstrual curses whereby the women's strategies become paradoxical because they may be interpreted as being prepoliticial. The second reason may be a case of epistemic ignorance as theorized by Charles Mills (2007). In 2015, Mills extended his conceptualization beyond the United States to analyze global inflections of white ignorance. In short, white ignorance indicates "an ignorance among whites—an absence of belief, a false belief, a set of false beliefs, a pervasively deforming outlook—that was not contingent but causally linked to their whiteness" (217).

The silence of the white men could be due to their incomprehension of the ritual. Despite the translation efforts marshalled by the organizers, could it be that the ritual cursing is too opaque—read, too symbolic—to make sense to the Parisian media and political establishment? Alternatively, is the lack of coverage an index of the demand for transparency from certain corners of the globe? Relatedly, is it the rejection of actions that refuse to conform to "global"—read, North Atlantic—standards of claims making? That their incomprehension results from the opacity of the ritual is the most hopeful scenario but also the least plausible in light of the centuries-long involvement of France in Côte d'Ivoire. With the ever-complex mobilities of our (post)modern world—where frontiers are expanding, and certain boundaries and borders are becoming increasingly porous, allowing more and more peoples and cultures to be cast into intense and immediate contact with each other—the need for transparency seems to have reached its pinnacle. This dire need works in tandem with the distrust and refusal of opacity in all its forms, including the language of condensational symbolism. So in this grammar of globalization, what space is made for people who still value the symbolic and the figurative? Perhaps globalization is much less about the constitution of the globe into webs of interconnection than it is about the increasing global standardization of certain practices, tastes, ideologies, and cultural goods, showing, if it is still necessary, that the experience of globalization is a rather uneven economic, political, and social process (Appadurai 2000; Hall 1996).

If the ritual still seemed opaque, despite the explanatory efforts by Topo and others, the "colonial library" on Ivorian society by a wide variety of French so-called experts (anthropologists, ethnographers, political scientists, geographers, civil servants, historians, journalists, and expatriates) could certainly provide answers to the officials, if need be. More specifically, numerous contestatory events involving cursing dances during the anticolonial-era struggle exist in French archives on Côte d'Ivoire, as I discussed earlier. However, given the changing nature of societies, certain kinds of action might change meanings, making an update on the meanings of the kodjo rouge more than necessary.

I read the involvement of the organizers of Opérations Kodjo Rouge in multiple demonstrations that preceded and followed their rallies as ways of framing their acts. For instance, in April 2011, a few months before the first cursing rally, between 1,300 (according to the police) and 3,000 (according to organizers) people demonstrated in Paris to denounce the military operation that overthrew the former Ivorian president Laurent Gbagbo (AFP 2011; Cantiniaux 2011). Almost a year after, on March 31, 2012, several Ivorian organizations in France demonstrated in front of the Elysée Palace in Paris to remember March–April 2011 (Togui 2012a).[22] This remembrance became the opportunity to demonstrate against Sarkozy and the contribution of the French army in that military offensive. Again, on July 13, 2012, a month before Opération Kodjo Rouge III, another coalition of Ivorian organizations gathered in front of the Palais Bourbon, the French National Assembly, and even submitted a motion to the president of the National Assembly, M. Claude Bartolone, protesting against French politicians' troubling interventions in Ivorian politics.[23]

Silence of officials may be because the white men understood the implications of cursing but simply, because of their cultural locatedness, lacked appropriate ways to respond to it. Again, this is highly debatable in light of the accumulated knowledge by the French civilizing mission in Côte d'Ivoire and the multiple acts of translation and contextualization by the organizers of the rallies. More specifically, the North Atlantic library suggests that a correct response to the women's cursing would be that the targets (Barack Obama, Nicolas Sarkozy, Abdoulaye Wade, and Alassane Dramane Ouattara) should make public statements acknowledging their mistakes and taking steps to make amends; in that case, the ritual participants would organize a propitiation ritual to reverse the curse.[24] Failure of the targets to demand forgiveness and make offerings to the ancestors will lead to their misfortune or death. However, for the different Opérations Kodjo Rouge, no public statements from the targets were issued, leading one to believe that the curses are still valid or that nothing bad had happened to the targets.

Finally, the silence of the French political class may be read as evidence of their disbelief in the power of menstrual rags, thereby turning their reaction into what I claim is an epistemology of ignorance. This more sinister interpretation establishes similarities between the fictional Commandant Kakatika and the current political class. Epistemic ignorance is a reformulation of Charles Mills's "Global White Ignorance" (2015), which he defines as structural forms of motivated irrationality, self-deception, and implicit bias that are causally linked to whiteness, understood as a political rather than biological construct (217). That ignorance should be understood not as a lack but instead as a successful structural practice that has maintained the dominant group since the Enlightenment. Epistemic ignorance opens up ways to account for how the silence of French public officials on the cursing rallies works as an oppressive practice, a practice marshalled to maintain the dominance of the privileged group. While the women, as part of the oppressed and the initially injured group, have fewer motivations in ignoring the sociopolitical order that caused the oppression in the first place, their targets have numerous interests in maintaining the status quo, even if that means disavowing one of the fundamental principles of their professed Enlightenment.

Enlisting the silent reaction of the women's targets enriches our account of the women's ability to secure the expected outcome of their cursing ritual. The unpredictability of these reactions derives from the limits of the body to speak universally despite biological sharedness. Clearly, what is considered naked is geographically, historically, and culturally context specific. Even in protest, we cannot assume that nakedness provides definite answers; in fact, it raises questions without necessarily communicating the actors' intentions.

Translating, Mediating, Modifying

For Opérations Kodjo Rouge to be visible on the global stage, the participants had to translate, mediate, and implement massive adaptations to conform to the sociopolitical realities of their chosen environment, Paris. These translations/mediations and modifications include deterritorializing, reterritorializing, internationalizing, degendering, breaking the opacity, clothing the ritual, and mobilizing social media and popular songs. Though utterly necessary to inscribe the women's organizing within the global account of resistance movements against neocolonialism, however, these adaptations and translations contribute to the misreading of the menstrual cursing as nonviolent.

The role of globalization as both enabling and disabling is even more compelling in relation to these cursing rituals. Deployment of the Kodjo Rouge in Paris speaks to one of the fundamental features of globalization: the increasing weakening of national sovereignty in previously colonized spaces and its allocation to former colonizing powers in the lightly disguised forms of multilateral organizations. These supranational and transnational institutions, such as the United Nations or the International Criminal Court, bodies that Michael Hardt and Antonio Negri (2000) name "empire," consistently use appeals to human rights as the basis for global governance.[25] Thus, Opérations Kodjo Rouge participants demonstrate their keen understanding of the location of political, cultural, and military power in the globalizing era in the way they make their claims in Paris.

In their work of translation in multiple media appearances and social media materials, the organizers extended the intended meanings of their act to the larger audience. As mentioned earlier, the uncritical use of generalizing categories such as "Africa" and "women" suggests that they were targeting a more general audience, one not necessarily familiar with the workings of their customary system. But some of these instances of translation lamentably failed, producing self-contradictions and thus garbling dramatically the projected meanings of the kodjo rouge. Similar instances include Zéka Togui's framing of the rallies as a peaceful mode of resistance using lethal weapons: "This will be a historic event in the history of peaceful resistance of Ivorians in the diaspora against neocolonialism and despotism in Côte d'Ivoire. It is in the coming together of indignation and revolt without knives, without Kalachnikov, without bombs. These women are determined to launch the offensive against Alassane Dramane Ouattara and Barack Obama. For this umpteenth offensive, she wielded the lethal weapon, the kodjo rouge Terminator" (Togui 2012c).[26]

In the same sentence uncritically coexist terms of different semantic fields: *peaceful resistance, offensive, lethal weapon, terminator,* and *kodjo rouge.* It appears that the blogger is aware of the possible contradiction between *peaceful resistance* and *lethal weapon/offensive.* The name of the rallies, Opérations Kodjo Rouge, and its heavy association with military interventions disrupt this inaccurate rhetoric of peace; further, the ritual participants looked to harm their targets. This instance of failed translation reflects a disjunction between members of the organizations and their bloggers about how they ought to frame their oppositional tactics. Topo uses a rhetoric of war making, precarity, and symbolic wounds, and of women being trapped in a state of injury and without options; the protest bloggers, on the other hand, refer to the rallies as "la résistance pacifique" (peaceful resistance) (Civox.net 2012; Togui 2012c). How does one reconcile the framing of the women's position as both a matter of life and death and a peaceful strategy of resistance? More importantly, by couching the women's ritual in the rhetoric of peace, the bloggers unwittingly silence the protesters' radical tactics in the name of translation and global unity, thus corrupting their message.

Further, the efforts of translation via images and videos of the rallies—which move at a higher speed, creating an ever-denser network of cultural and political interconnections—introduce another layer of self-contradiction. The images and videos show the festive atmosphere of the rallies, during which women pose for the camera in provocative and sexual positions, contradicting the grave atmosphere described by both their leaders and bloggers. While this type of circulation provides the women an international stage for voicing their grievances, it also contradicts the narratives of cursing and the production of death that the organizers hoped to communicate. Indeed, the festive mood conflicts with Topo's words of injury: "Ivorians are hurting because of the high death toll caused by the bombing of the Residence of Laurent Gbagbo and of Ivorians, under the order of Nicolas Sarkozy" (Togui 2012b).[27] Upon first reading Topo's interview, I seriously believed in the gravity of the women's threat to curse. However, I was simultaneously shocked and amused to see some videos of the events that then suggested a festive party. The wider circulation of images thus becomes a double-edged sword, reinforcing but often putting pressure on the more constructed and controlled images produced during interviews and news reports. In this performance of assaultive resistance and vulnerability, globalization plays both enabling and disabling functions. Globalization made available to most located outside of Paris news and images of Opérations Kodjo Rouge, thus allowing widespread participation in the imagined and real actualities and struggles of cultures from around the world.

To communicate effectively the women's goals, their naked exposure was enhanced or supplemented by nonverbal and verbal communications that they disseminated through various social media platforms. During the performance, their body language includes making hand gestures of imploration and/or defiance: groping genitals, holding breasts, pointing to bellies and buttocks. The meanings of the nonverbal communication are subject to misreadings. The women have as much agency as their targets, who may choose to impose their self-serving meanings onto the gestures. Therefore, the women's body language proves insufficient and needs to be complemented by verbal communication and written signs. More often than not, the women need songs, chants, incantations, and curses to complement, clarify, and even to strengthen the symbolic meanings of bared body parts, especially when appealing to supranational bodies. These may also show their limits because they need translation and mediation.[28] These words publicly expose the women's grievances, list their targets' offenses, and create a unifying emotional connection between them.

Out of hundreds of naked protests organized in the last twenty years, most have included visual signs. Opérations Kodjo Rouge built on the archive of previous social movements by adding posters, placards, T-shirts, and other signs. The nationalistic atmosphere was expressed through T-shirts with the name *Côte d'Ivoire*, or pictures and/or the name of Laurent Gbagbo, or messages such as "Gbagbo: the choice of the people, of African sovereignty and dignity." In this language, one sees the expansion of Gbagbo's image as the liberator of the Ivorian people to that of representative of an oppressed black consciousness more generally. Several posters show a picture of Nicolas Sarkozy, flipped with his head down, accompanied by the injunction in French to "back off." Such images are sometimes accompanied with the universal symbol of death, the skull and crossbones, and the call "Women of the World, Come Join Us to Bust Out Sarkozy."

One poster shows images of previous demonstrations complete with the telephone numbers of the main organizers to contact with questions (figure 7.3). By providing names, banners, organization tables, and leadership information, these women and their allies deploy what Charles Tilly (2008) has identified as the recurrent and predictable "toolkit" of protest tactics used by collective actors globally to make claims, promote goals, demand recognition, and convey numerical strength.

However, even these written texts present challenges despite or because of the encroaching standardization of cultural and economic globalization. Given the ethnic diversification and the existence of multiple indigenous languages around the world, witnesses have to rely on translations by reporters, who in turn also may need to rely on translations from bystanders, leaders, or other mediating figures. Although their written signs suggest the limits of symbolic messages, they also demonstrate the women's keen understanding of the unstable meanings of their acts beyond their immediate contexts and their desire to control somewhat the multiple interpretations that they may be subject to. Without these, we often get exposed bodies and inaudible voices. When women add their voices to naked protests, their agency is harder to deny. Whether clothed or not, they are no longer merely objects of anyone's gaze. As such, attempting to conceptualize defiant nakedness is fraught. Although difficult, it is utterly necessary to decode the texture of women's agency.

Although the deterritorialization and reterritorialization of the kodjo rouge shows the women's awareness of global loci of political and military power, they also require adapting the ritual to the prevailing social norms, cultural contexts, and laws of Paris. Specifically, the organizers had to innovate within limits set by the French penal code and its Article 222-32 prohibiting nudity and seminudity in public settings, which stipulates, "An indecent sexual exposure imposed on the view of others in a public place is punished by one year's imprisonment and a fine of €15,000."[29] Certainly, there is a dire need to abide by the local laws for the purposes of the demonstration permit and to avoid systematic arrests by the police force. In May 2014, Topo Léontine herself showed her awareness of that law, inviting women to wear white outfits because "here we are in France and can't go naked" (Eventnewstv Presse 2014). Independently of that legal constraint, I suggest there is something else involved in clothing the ritual with white pants, leggings, shorts, cropped pants, and T-shirts. The reason is that exhibiting the participants' physical nakedness comes with their possible instrumentalization by social media and the press. Even more importantly, the

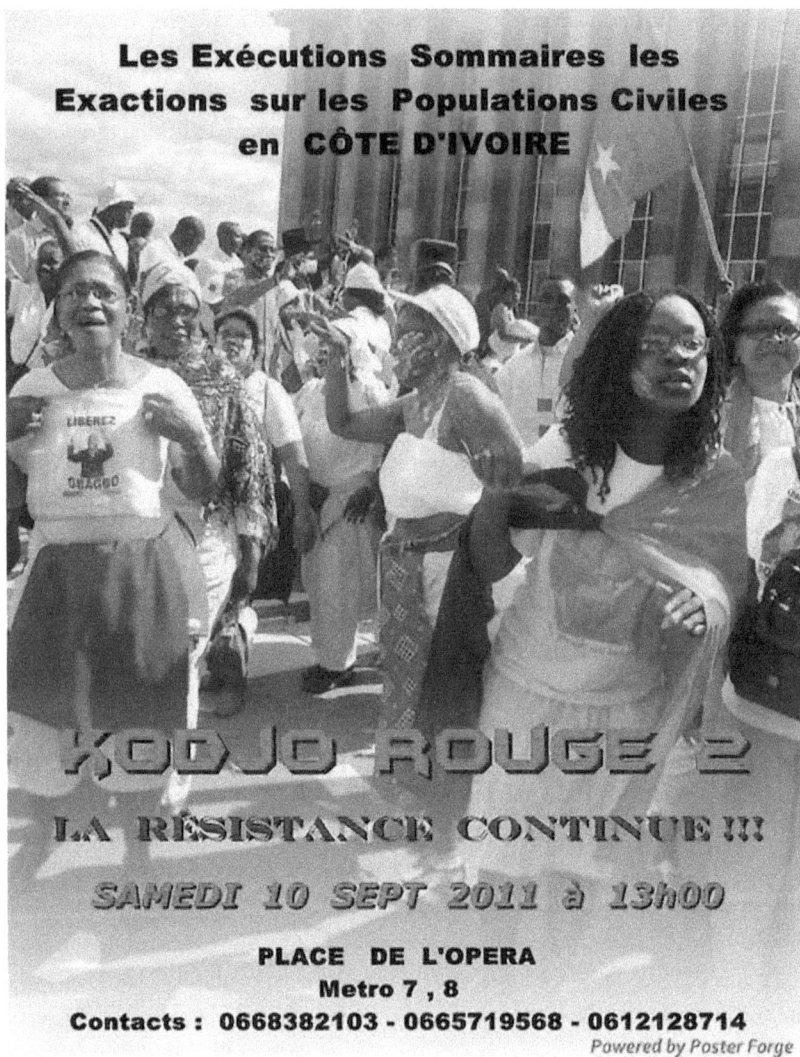

FIGURE 7.3   Poster for Opération Kodjo Rouge II, March 2012.
Source: *La Dépêche d'Abidjan.*

presence of cameras that record, disseminate, and preserve for posterity comes with the possibility of the women being haunted by the images of their nakedness, as the women of Dobsonville have shown.

On the question of whether the clean, red cloth that the women wore over their pants would be effective in materializing their curses, Topo answers that the cloths were previously used during menstruation but were now clean. Of course, the organizer's statement remains questionable given the similarities between the cloths—as if they were "cut from the same cloth," no pun intended.

In addition to clothing the ritual, the organizers also degendered and internationalized it. Regarding men and their participation in what is conventionally an exclusively female ritual, Topo provides no rationale. Rather, she assures her interviewer (Togui 2014) that men will be protected from the curse with a red headband (their kodjo rouge) and a special white clay that will be available at the rally. This precaution is necessary because in other settings, the scene of cursing with menstrual cloths is a fundamentally female-centered practice and men would systematically run away since being touched by the cloth is itself a curse. Inclusion of men as ritual participants is a major adaptation that requires further elaboration. In a world shaped by detraditionalization and globalization, "making the world in reality and in experience more interconnected" (Hall 1996, 619), modes of punishment, for example the curse of the kodjo rouge, are increasingly occurring outside kinship and family networks, and in societies that are becoming increasingly diverse. Given the women's acute awareness of these changes and taking advantage of the availability of men to participate in their ritual, they tweak the gender restriction with the goal of conveying numerical strength. Yet the participation of men to signal numerical strength may weaken the ritual's real and/or symbolic effects.[30] I suggest that by including men in the ritual, the women's primary goal may not be actual "cursing" but rather attracting the attention of journalists and those unfamiliar with the Ivorian political landscape.

The organizers' courage in expressing their grievances in ways that reflect their worldviews is remarkable. With the employment of modern-day conventional resource-mobilization tactics, such as Facebook and YouTube, in combination with menstrual cloths and calling on ancestors to curse, these women fashion new ways of being in and around the world. However, their modes of resistance are paradoxical because genital cursing may reinforce the view of women in Africa as primitive and backward. Although these cursing rituals reflect subalternity and its overdetermined pronouncements, the ritual participants enrich the conversations on resistance movements in the international arena. Their contribution consists in highlighting a mode of organizing wherein invisible agents matter. They thus offer the opportunity to reflect on conceptualizations of agency that are more capacious and that account for social change with the assistance of gods and spirits. If in colonial India, such oppositional politics were at best framed as prepolitical, what terms will accurately describe these women, whose mixed strategies belie the backwardness of their undertaking?

If attention to social media in analyzing genital cursing is productive as it helps to decenter a narrow reading of the profile of women who disrobe for political purposes, literary fiction plays a similar role. In closely reading Echewa's historical novel *I Saw the Sky Catch Fire* ([1992] 1993), I account for the murderous responses that self-exposure met in the colonial period to argue that the novel displaces the commonplace images of victimhood attached to women's bodies without falling into the easy trap of triumphalism.

(Mis)Reading Murderous Reactions

The White Officer, Malcolm Davis, stepped forward toward the women, flanked by two soldiers and an interpreter. He held a piece of paper in his right hand, ready to read the women the riot act. "Ndom kwenu" Ugbala intoned. "Huo-ra-nu Nwa-Beke-e ikeh! Huo-ra-ya-nu otila!" [Show the White man your backsides!] As if on cue, all the women turned around, facing away from Mr. Davis and his soldiers, doubled over, turned up their buttocks and aimed them at the approaching White man. Then they pulled up their loincloths, so as to expose their naked bottoms. Mr. Davis did a double take, then froze in place. The soldiers with him turned away their faces, as did the formations behind them. It was a sight none of them had ever seen before and hoped never to see again. They felt insulted, assaulted, defiled, and cursed. —**T. Obinkaram Echewa**, I Saw the Sky Catch Fire

The passage quoted here is from the most dramatic scene in T. Obinkaram Echewa's historical novel *I Saw the Sky Catch Fire* ([1992] 1993). The novel includes a fictionalization of the 1929 Igbo Women's War, during which thousands of women opposed colonial incursion and exploitation. They deployed multiple contestation strategies and even showed their naked buttocks. Following the historic exhibition, the forces of colonial administration shot dead more than one hundred women. This female public disrobing event was one in the long series of women's contributions to struggles for independence throughout sub-Saharan Africa.

The novel demonstrates the powerful necessity of including reactions to women's cursing self-exposure in the conceptualization of their agency. As previous scenes have shown, insurgent disrobing is a performance that targets audiences whose responses constitute an essential element of a more comprehensive account of their tactics. These reactions—often immediate, often differing, but always continuously unfolding—represent the naked and open aspect of "naked agency."

The Igbo women's historic collective act, aimed at the representative of the colonial administration, is fraught with tension given the multiple and often contradictory responses that it enables. Assuming the invisibility of the African soldiers, the proxies of the colonial administration, in Ugbala's command as she singles out the white man as the target of women's genital cursing ("Show the White man your backsides!" [208]), is fatal. She underestimates the murderous capacities of the soldiers, who understand the curse of female exposure and who, Ugbala thinks, would not shoot at their "mothers." The entity that ends up causing damage was not considered as a target, suggesting the necessity to consider all possible feedback, near and far.

I number at least four reactions in accordance with four entities: Igbo men, the colonial administration, the *munchi* soldiers (after the Munchi plateau districts), and scholars. The first three reactions are suggested in the novel, and the fourth is regarding the novel. First, although not the direct targets of the women's assaultive exhibition, the soldiers, who are from a neighboring region of Igboland, know the "intended" message of the exposed buttocks and react with murderous violence. Second, the Igbo men also understand the dangerous implications of mature female insurgent self-exposure as they flee to hide. Third, the colonial administration, the target of the act, dismisses the women's act as the manifestation of insanity and mysticism. The fourth reaction, which forms part of the continuously unfolding response, emerges from the field of African studies. Previous analyses of the novel and contexts of the Women's War provide triumphant narratives of heroism, ignoring some agents, the soldiers, and the paradoxes inherent in this kind of contestatory act. The binary and dualistic thinking (the women versus the colonial administrator) in Ugbala's injunction— instead of the triad of women, soldiers, and colonialists—is symptomatic of previous readings of the novel.

In this scene, I ask, What is the role of literary fiction in our current understanding of resistant genital flashing? I answer this question by uncovering another aspect of women's agency: their inability to foresee and control the responses of their targets and those of bystanders. *I Saw the Sky Catch Fire*, one of the most iconic representations of naked protest in Africa, deepens our understanding of naked shaming by offering multiple perspectives that are unsanitized and unsubordinated to ideological investments, including overrepresentation of elite documents among them. Through the 1929 Women's War, the novel brings to light an affective register, the protesters' as well as the targets' feelings and value judgments, which is unavailable in journalistic sources. Departing from readings that see the scene of collective genital cursing during the war as dyadic, I highlight the presence of soldiers who understand the gravity of the act and therefore kill more than one hundred women to prevent them from extending the curse.

In the novel, the characters' emotions reveal the unexpected and counterproductive effects of uncivil self-exposure. With its abilities to give the reader access to the experience and interiority of characters, the novel differs markedly from the condensed nature of, the instantaneity of, and the sensationalism inherent in most news reports on defiant nakedness. These elements of length and exposition make the novel a vital art form with which to draw a new conceptual map of protest nudity, enabling a nuanced method for analyzing naked protest, which has thus far received little attention from literary scholars.

The 1929 Women's War

The Nigerian-born and U.S.-based scholar Echewa published his historical novel in the United States in 1992.[1] For an African novel, *I Saw the Sky Catch Fire* received an unusual amount of attention in the American and English print media. Within a year of its publication, enthusiastic reviews appeared in about thirty English-language newspapers and magazines. The main trend in these reviews concerns the women's "unorthodox"—in the dominant Western sense—retributive strategy in an asymmetric dynamic of power, opposing rural and colonized Igbo women to the British colonial enterprise with its economic and military privileges. Among scholars, however, the novel has met with a relatively cold response; the first critical essay did not appear until 2000, eight years after the publication of the novel.

The novel includes a fictional reconstruction of an actual event, the 1929 Women's War, which was women's aggressive reaction against the colonial incursion in Igboland that resulted in taxation on women for their belongings, including goats, chickens, and palm oil. *I Saw the Sky Catch Fire* masterfully blends history, tales, and anecdotes, blurring the conventional assumptions around transmitting women's stories. Divided into two major parts, the retrospective story is told from the point of view of an overarching first-person narrator, Ajuzia, who went to the United States to pursue a graduate degree, returned after five years to bury his grandmother, and found his wife pregnant by another man. In the first part, set in a village at the peak of colonialism, Ajuzia relates his nonliterate grandmother's stories of women's courage before and during the famous Igbo Women's War of 1929. On the eve of Ajuzia's departure to the United States, Nne-nne tells the stories to remind her grandson of her ability to run the family compound while he is away in the United States. The grandmother's stories highlight how Oha Ndom, the Igbo name for "women's solidarity," creates the powerful and sacred bond that unites women and provides the leverage they enjoy. One of Oha Ndom's most memorable expressions of power is women's use of social nudity in their fight against local sexist practices and the colonial incursion. The second half of the novel recounts Ajuzia's life narrative and his struggles as a postcolonial intellectual torn between pursuing his dreams of defining himself and fulfilling his obligations to his family. Convinced of his grandmother's ability, Ajuzia leaves to continue his education. Before dying, however, Nne-nne uses her charisma, and the prerogatives of an elderly woman, to save her grandson's marriage. The novel ends with the protagonist's decision to stay in Nigeria rather than return to the United States to complete his doctorate.

As a fictional narrative, *I Saw the Sky Catch Fire* also provides a new path through which to meet and fully appreciate women who dared bare their buttocks against white colonialists and their native allies. Officially called *warrant chiefs*, these native allies created their own specific forms of despotism, becoming figures that Mahmood Mamdani finds instrumental in the implementation of the "bifurcated state" (1996, 27). The Women's War of 1929, Ogu Umunwanyin in Igbo, became the first recorded female revolt against imperial forces in Igboland. It, alongside several collective female public disrobings, problematized the long-standing image of women in Africa as eternal victims of local and foreign versions of patriarchy. Other events include the 1933 Togolese women's purification rituals against the arrest of local leaders, and to contest an increase in taxes (Lawrance 2003). In the 1947 "vengeance of the vagina head" in Yorubaland, women of Abeokuta protested against taxes applied to goods and related issues in the marketplace (Johnson-Odim and Mba 1997); in the 1949 cursing/purification naked dance (Adjanou), as discussed earlier in this book, Ivorian female secret societies opposed colonial repression. The 1958 mass protests of Cameroonian women's indigenous institutions, called *Anlu*, and the 1959 beer hall riots and boycott in South Africa are part of this genealogy of collective action, as already argued.[2] Several historians and anthropologists have explored the significance of these struggles and their importance in localized resistance to colonial regimes of power (N. Achebe 2005; Bastian 2002; Johnson-Odim and Mba 1997).

The Women's War is essential to creating an accurate genealogy of literary history in African studies. For instance, Rhonda Cobham-Sander (2003) has challenged that conventional male-biased literary genealogy, placing the war in the tradition that gives us *Things Fall Apart* (C. Achebe 1958). Susan Andrade (2002) has claimed that the war stands as a symbol of a women's literary tradition and that uprisings like it are one form of visible political activism by women, against which we can read early middle-class women novelists as entrants to the public sphere.

Historically, the war is attributed to the British colonial administration's 1906 Native Revenue Proclamation, which founded a taxation system in line with the imperial philosophy that the colonized should bear the costs of colonization. Initially, the resultant taxation was applied only to men. Women were exempt from taxation because they were positioned under men and not recognized as full subjects to be subdued by the colonial administration.[3] Adult males were counted and taxed, and their failure to pay resulted in imprisonment (Afigbo 1982; Mamdani 1996). Taxation was oppressive, and the counting of human beings—necessary for assessing taxes—was always dreaded by the natives, who considered census taking as "provoking evil spirits, and causing destruction and death" (Afigbo 1982, 94).

Details about the war given in the novel are similar to the ones provided in historical studies. Nne-nne attributes the war to two main causes: the pressure of economically exploitative measures such as taxation and trade policies of the colonial administration, and the Igbo men's inability to contest them: "The Women's War started because the men did not start a war when they were counted and tax was imposed on their heads, . . . The war started because that was the third or fourth bad year in a row for everyone, a year of hardship during which palm trees bore little fruit and the fruit they bore produced little oil, and at the market the oil fetched next to nothing" (36). Factors behind the Women's War include questions of gender, drought, and economic interests. In relation to gender expectations, males failed to uphold their culturally sanctioned roles as the protectors of the community compounds. The dire economic circumstances, a result of the global economic downturn due in part to World War I, hit women hard, compromising their major financial independence and ability to care for their families.

Attention to the exposition of these contributing factors indicates the seriousness of women's grievances, as well as the meticulous nature of their organizing for the war, including devising sophisticated communication networks. Since the women were discreet in their planning, Igbo men as well as the colonial administration were unaware of and unprepared for their destructive powers. As the war escalated, the "White Officer," Malcolm Davis, entered the scene, declaring the riot act. However, rather than diffuse the situation, his action prompted the collective self-exposure, as narrated above. In addition to the shaming and cursing acts, the women damaged several native administrative centers, courthouses, mission houses, prisons, post offices, factories, roads, and other markers of European invasion of Igboland. Igbo men and soldiers understood the gravity of the act, whereas British colonialists silenced the anticolonial demonstrations by using large numbers of police, soldiers, and even Boy Scouts from neighboring regions. Police and soldiers killed about fifty women and one man. These figures, reported by the colonial investigative study, have since been contested.[4] The narrator of Echewa's novel gives the number of women killed as more than one hundred.

Although the war had far-reaching consequences, colonization continued in Nigeria until 1960. In the novel, however, women refuse to consider the violent encounter a defeat as they celebrate their victory and abilities to climb "the sky and set the sky on fire" (39). In acknowledging women's courage, the novel proposes an enabling narrative in opposition to the disabling narratives of British colonialists, without fundamentally falling into the problematic trap of erecting exposed buttocks as a noncontroversial form of political participation. Thus, *I Saw the Sky Catch Fire* retroactively anticipates Achille Mbembe's (2002) call to African intellectuals to take up the challenge of "African modes of self-writing," which would restore agency to Africans. The novel transcends the simplistic binaries of the narrative of loss/shame versus the narrative of agency. The novel also recognizes the very variables of gender and means of contestation, which are absent from Mbembe's article. Consequently, *I Saw the Sky Catch Fire* fits into a larger revisionist project that scholars of women in Africa inaugurated in the 1980s as part of the international rise of women's, gender, and postcolonial studies that began in the 1960s. These scholars, including Nina Emma Mba (1982) and Molara Ogundipe-Leslie (1994), have called attention to these otherwise obfuscated aspects in the historiography of African struggles for decolonization.

Effects and Affects

As a gendered and standardized form of retribution, the deployment of female self-exposure is predicted in the violation of normative codes against wife battery, demeaning postmenopausal womanhood, or perpetuating other egregious but unspecified acts. Whether in the form of shaming or cursing, which are often indistinguishable, the damaging effects of the act on both women and their male targets are devastating and even murderous. The effects of cursing and shaming for most involved explain how the motivation to resort to it reflects the lack of options in extreme circumstances. These fascinating gestures, often called "sitting on" or "making war" with an abusive Igbo man or with the colonial administration, even more powerfully reveal the social implications of aggressive nakedness in Igboland. These acts also speak to the importance of the abstract moral code of shame and its material effect as social death. Given the socially sanctioned nature of insurgent disrobing as a conflict management strategy, one would expect the performers to benefit from the indiscriminate compliance of the women's targets. If that were the case, women's agency would remain absolute and unquestionable. However, like all human actions, despite the force of legal, divine, or customary structuring mechanisms, the women's targets possess some level of agency. The existence of that agency, regardless if illegitimate or antisocial, encroaches on the absoluteness of women's authority and power—rendering the women's agency open: it has to be open; it has to be naked as it works with the agency of targets far or near, immediate or deferred.

I Saw the Sky Catch Fire is encyclopedic in terms of the multiple instances of defiant self-exposure that it narrates. It is also enlightening in showing how multiple categories of men or collectivities of men react to being unwittingly exposed to women's nakedness. Specifically, it provides an array of violent and counterproductive responses from women's targets or nontargets, as revealed by Ugbala's injunction during the war. Although the soldiers did not figure in her "show the White man your backsides," their affective reaction nonetheless inscribes the fateful nature of defiant mature female nakedness (208). Accessing the soldiers' interiority, the narrator exposes their feelings for being forcefully subjected to women's intentional nakedness: "Cursed by this field of female bottoms, fat and lean, old and young, brown and black, and at every stage of the monthly cycle. It was enough to give one recurrent nightmares and bad luck for all of a lifetime. The soldiers shut their eyes fiercely in the manner of little children learning to wink, as if thereby to squeeze the image of what they had seen out of their minds. They felt like throwing away their weapons and running" (208).

As the soldiers attempt to make sense of what they have just experienced, the women charge them with a cry, "Shoot your mothers! Shoot your mothers!" (209)—and the soldiers do, killing more than one hundred women. It remains unclear why the soldiers shoot. Did the killing result from the women charging or is it because the soldiers felt cursed? Unfortunately, for the women, "reveal tabooed body parts to defeat" instead turns into "disrobe and die." Within the soldiers' superstitious feelings reside the seeds of death—not of themselves but of the women. The killing of the doubly mothering genitalia resonates with the teleology of cultural progress; the colonial ideology, in which the soldiers are puppets, sees moving from the feminine toward the masculine as progress: going from a lower to a higher order, from mysticism to political rationality. The anxiety about *not* moving forward is at the root of so much violence on the part of the soldiers.

The steps leading up to the women exposing their genitals are described dramatically. For instance, in describing the women, the series of phrasal verbs "turned around," "doubled over," "turned up," and "aimed at" communicate a sense of urgency that enriches the reader's understanding of the scene. These details work like a linguistic video of the cursing, ensuring that the significance of the women's act is not lost on the reader. It is precisely these details that set the novel apart from historical, anthropological, and journalistic texts, where we seldom have access to the targets' feelings. More generally, access to the targets' as well as the actors' feelings and emotions makes the novel unique in that it enables the appreciation and analysis of the effects of naked cursing, including its counterproductive and unexpected consequences. Through the collocation of adjectives interspersed with commas, "felt insulted, assaulted, defiled, and cursed," the passage heightens the feelings of destruction through restatement, expansion, and intensification. The pauses materialized by the commas slow down the otherwise fleeting and fast-moving event. They render the women's gesture more memorable by stamping it on the reader's memory.

Narrative distance is also manipulated in relation to accessing the characters' interiority. From the first-person narrator with its de facto limited access to other characters' feelings, the reader then encounters an omniscient narrator's unrestrained access to them. The manipulation of narrative distance circumvents the limited perspective of the first-person narrative.

Although not the targets of the protesting women's retributive strategy, Igbo men reacted according to the women's expectations when they hid from the women's acts. The prevalent Igbo cosmology regarding the powerful social institution of motherhood, the sacredness of women's solidarity, and the privileges of seniority was not lost on men who upheld it. Women's socially elevated position over men's is meant to communicate the potency of mature women's potential retaliation with their bodies. Nne-nne, the grandmother in the novel who functions as the repository of customary practices, explains to her grandson how anatomy is the site that hierarchically categorizes men and women: "Ajuziogu men and women are like their organs. A woman's is mostly private, tucked away like a secret purse between her legs, with little to give away how big or deep it really is. A man's on the other hand, hangs loosely and swings freely about for all to see. A few years into old age, and men have to offer sacrifices and pour libations for their erection" (10).

Women and men resemble their genitals and therefore are constituted in a parallel mode to one another. In this binary thinking, the ovaries and the testicles saturate the whole being and determine men's or women's behaviors, a view that is similar to the one held in nineteenth-century Europe (Laqueur 1990; Moi 2001). In the novel, the parallel structure mutates into hierarchy, which is reflective in the intrinsic qualities of genital organs: hardness and limpness, openness and closure. A historical force such as colonialism reinforces the hierarchy and further reifies the gap between men and women. Men's openness renders them porous, and their limpness shows their inability to fight against colonial intruders. Contrary to men's genitals, women's genitals are understood as "secret purse[s]," an economic metaphor for women's contribution to the community's protection and survival. Thus, women's bodies create awe in men, a social belief that Nne-nne aptly underwrites: "A woman is everything! . . . A woman is like a god! A woman's crotch is a juju shrine before which men always kneel and worship. It is their door into this world" (14). In the case of genital cursing, it is also their potential exit. This belief in the power of women is aptly expressed through the Obie N'Ole?, the crotch dance, which women in the village sing whenever a baby is born:

Tall men?
From the woman's crotch!
Short men?
From the woman's crotch!
Hunters and warriors?
From the woman's crotch!
Chiefs and court clerks?
From the woman's crotch!
Even the White man?
From the woman's crotch! (14–15)

Inclusion of "the White man" in this list is an assertion that this cosmology thinks of itself as universal, or at least transferrable to white westerners. In the novel, the inclusion of Oha Ndom's songs such as the crotch dance responds to the need to convey the far-reaching implication of this communal record-keeping form.

Whereas the nontargets' response was materially murderous, the "White" men's response was ideologically disempowering for the women: "Insane! Irrational, Mass hysteria, like the spirit-induced madness that possesses some of them during some of the juju festivals! A sudden overflow of premenstrual or postpartum hormones! Spontaneous Combustion" (39) are some of the terms that the colonial administration used for decades in response to the women's gesture. This response—highly predicated on primitivism, pathology, and anatomical determinism—shows a markedly different understanding of defiant disrobing. The dismissive response is not only gendered with the idea of hormones but also highlights the categories of fetishism and hysteria and the denial of empirical reason that the colonial administration considers absent from the natives. The colonial response resonates with the one that the performers of Opération Kodjo Rouge recently encountered in Paris.

With their rhetoric of excess and of uncontrollability combined with that of the intervention of dark spirits, the "White" men dismiss the meticulousness of women's organizing. This dismissive move is a political and ideological maneuver to deny women the consciousness and the organizational schema for punishment and protest. Even the naming of the events as "riots" in the official history of the British downplayed their significance (Van Allen 1976).[5]

In addition to the strategic failure to read the war accurately, the historical constructions in racist colonialist texts of trivialized African women's naked bodies contribute to explain the gendered and racialized reaction. "White Officer" Davis's surprise stems not just from his unwitting exposure to Igbo women's nakedness but also from the circumstances of the act. Unlike the objects in nineteenth-century missionary ethnographies, here protesting women become the agents of their bodies and actions. In the nineteenth century, in attempts to document the unrestrained nature of colonized African female sexuality, the voyeuristic eyes of colonial "photographers" metaphorically undressed and trivialized women's bodies, setting them in opposition to the untouchable body of the British middle- and upper-class woman, the "damsel in distress" (Corbey 1988; Gilman 1985; McClintock 1995). Encumbered by their ideological baggage in the novel, the colonial administration refuses to recognize the colonized Igbo female body as either speaking or rescuable; to the administration, these bodies need no rescue because they escape the script of what has come to be framed as "the cult of true womanhood" with features of "piety, purity, submissiveness, and domesticity" (Welter 1966, 152).

So, although women disrupted the colonialist paradigm by inscribing agency with their bodies, they could not be read because they were buried under compounding layers of subjugating discourses. Through their daily drudgeries of laboring under the unfailing sun and the unbearable heat, Nne-nne and other female characters exemplify the "officially unprized and unguarded" bodies (Bennett and Dickerson 2001, 2). Just as enslaved females were masculinized in the slave economy, so too were the Igbo women in the novel; they undergo a discursive masculinization in the eyes of the British colonial administration, becoming exploitable and disposable bodies. However, the women proved wrong that political calculation; using their grassroots leadership and sophisticated networks of communication, they held meetings, raised funds, planned their attacks, and sent delegates throughout Igboland to mobilize other women in an elaborate, effective, and discreet way (N. Achebe 2005; Byfield 2002; Okonjo 1976).

I Saw the Sky Catch Fire conveys women's extensive preparation for the battle, describing women as being outfitted in ways that signify war in Igboland: wearing loincloths; carrying sticks wreathed with palm fronds; smearing their faces with indigo, charcoal, or ashes; and binding their heads with young ferns. The smearing of faces with indigo is an attempt to renounce their personal identities and their social roles as mothers, daughters, or wives. By stripping themselves of individual social identities in order to espouse a depersonalized and undifferentiated identity, the women sought to let the female principle stand out; at that particular moment, it is the only identity worth fighting for. In war attire, women were unwifely and militaristic. The short loincloths indicate their readiness to undress. The sticks they carried, considered more powerful than the guns of their oppressors, were saturated with symbolic power as they invoked the power of female ancestors (Harry 1970; Meek 1937; Perham 1937).[6] Demonstrations at the offending sites consisted of singing, dancing, and hurling verbal abuse at their targets.

As a performance overwhelmingly predicated upon a specific desired response from the targets (shame and guilt), Igbo women's targeting the British colonial administration with their historic exposure raises many questions: How do colonized women stage a protest using their defiant exhibition without reinforcing "epistemic ignorance" readings of primitivism? Would not the meaning of public undressing be lost on men outside the given culture? As I have demonstrated in previous scenes, especially with the use of stained menstrual cloths in Paris, Igbo women's agency cannot be understood as absolute; rather, it is a "naked" agency because it constitutes agency that is both contradictory and temporary. The constantly fluctuating nature of cultures shows the continuous necessity to redefine the meanings of exposed body parts, revealing faults in a given system of beliefs. Because of the customary nature of defiant disrobing, because the effectiveness of the gesture depends on a singular mode of recognition, and because the colonial administration possesses the agency to recall white epistemic ignorance, the women's defiant disrobing is subject to misreading, especially when deployed against colonial authorities. Even if the gesture is not subject to misreading, its effectiveness can backfire, as was the case with the soldiers.

The richness of *I Saw the Sky Catch Fire* in terms of multiple instances of self-disrobing invites further analysis in order to understand how various men or male entities react after being exposed to women's nakedness. In addition to the soldiers, the colonial administration, and law-abiding Igbo men during the war are Nne-nne's husband's relatives and an alienated abusive man during nonwar times. The first pages of the novel focus on the confrontation during a witch trial between Nne-nne and her husband's enemies and relatives, which sets the stage for the kind of counterproductive reactions to women's ritual nakedness that ought to be considered in conceptualizing their agency. The inability of the grandmother's husband, Nna-nna—described as "a nerve-grating complainer"—to fight his fellow men forces Nne-nne to step in and fight for him. Ajuzia recalls the incident: "Nne-nne had pulled up her loincloth and turned her naked buttocks up into the faces of the men's assembly and dared them to do their worst with her!" (10). The men were baffled by the woman's dangerous act, but they avenged themselves on her husband. This damaging and unexpected effect highlights the limits of Nne-nne's agency. Of course, according to customs and the moral guardianship bestowed upon mature women, her disrobing is powerful in that she used it to mark the potency of her social position. However, although empowered, her loved ones, specifically her husband, are left vulnerable to the backlash. In this case, the backlash of defiant disrobing targets the woman in an oblique way, as it harms her husband. Could Nne-nne's assaultive gesture be considered powerful? It is momentarily powerful because it elicits a potent response. However, it is not the feedback that Nne-nne expected. Her gesture and its effects remind us of the problematic rhetoric of triumphalism that has been brought to bear indiscriminately upon women's insurrectionary nakedness.

Besides the war, one of the most memorable events in the novel is the women's punishment of villager Ozurumba for beating his pregnant and helpless wife, Ekpa-Ego. In the village, it is common knowledge that Ozurumba is infertile; after years of marriage and childlessness, his wife takes a lover, who impregnates her. Upon learning of her infidelity and pregnancy, Ozurumba beats her, giving her a black eye. Following the lead of Ahunze, the widow and economically successful woman—known disparagingly by the village men as the "impossible wife"—whom Ekpa-Ego has befriended, the market women condemn Ozurumba's act and decide to teach him a collective lesson by using genital shaming. Organized with a plan of action, they take siege of his compound, publicly expose his offenses through songs, strip him naked, brandish his loincloth as a trophy, and urinate in his compound—turning it into a metaphorical no-man's land. Then, they use obscenities, and push him "to the ground and spread him out, face up, while holding his hands and legs so he could not struggle free. Finally, they take turns sitting on him, pulling up their cloths and kirtles to bare their buttocks and planting their nakedness on every exposed element of his body" (146).[7] Sitting on Ozurumba also suggests transferring the stigma associated with female genitals to his body.[8] Rather than its opposite in conventional accounts, here stigma is turned into a site of self-empowerment because women mobilize it for retributive purposes. Given the predictable nature of the punishment, male Igbo characters refrain from physically disrupting the collective punishment of Ozurumba. Furthermore, they feel helpless. Although they seek to prevent the punitive sanction through verbal negotiations, they fail.

The vivid descriptions of actions by the village women are meant to heighten the offender's feelings of humiliation and give the reader as comprehensive an understanding of the punishment as possible. The angry women plant their naked buttocks on Ozurumba's naked body. The figurative term *planting* is particularly revealing; it expresses the women's determination and potential for retributive violence, as well as their goal of putting onto Ozurumba's body the seeds of shame from the "soiling" genitals. The significance of the potent effect of sitting on a man is not lost on Ozurumba, as the narrator explains: "Ozurumba felt severely aggrieved. . . . A disgrace beyond repetition has been visited upon him, the ignominy and shame of a lifetime. [Women's] smelly and sweaty bottoms on his face, their rank wetness and the prickly hair of their things in his mouth and eyes! Unspeakable abomination. He would have rather fallen into a latrine" (151–52). The collective punishment leaves Ozurumba dazed, unable to articulate linguistically the effects of the most humiliating event that has befallen him. According to the narrator, Ozurumba referred to women's genitals as "hairy" and "smelly." These descriptors reflect Ozurumba's goal of metaphorically consigning those who had shamed him to the natural and untrammeled world of animality. We see, based on his emotions, an association between hairiness, bestiality, lack of hygiene, and lack of civilization (Ableman 1984; Barcan 2004; Basow and Braman 1998; Shaman 1996; Toerien and Wilkinson 2003).

Eja-Egbu, one of the village's oldest men, frames the incident as an abomination and notes its near-extinction: "Women used to sit on people a long time ago, but I have not heard of it in recent times" (149). Although Eja-Egbu does not explain the infrequency of the practice, it seems that colonialism, with its introduction of Christian values and mission schools, contributed to driving it underground. In the village, it is considered a disgrace of the highest order for a man to experience the "sitting on." For him to be stripped naked and sat on by a female, or several females, amounts to being in the bottom position, a place conventionally reserved for the female body in a man/woman pair. It is the equivalent of being "ridden" like a donkey, and therefore being denied authority in the household and, by extension, in the village. Thus, Ozurumba is literally dominated and symbolically castrated, resulting in his degraded social status. This hard treatment is the women's way of communicating to him the seriousness of his offense. The sheer number of women, about fifty, who sit on Ozurumba causes him to relive repeatedly the haunting experience of symbolic castration.

An action that started as "banal" domestic violence, viewed differently by men and women, escalates to the gruesome murder of the woman whom Ozurumba accuses of being the root cause of the punishment, as well as his suicide. Because he was taught a lesson and symbolically castrated during the genital shaming, Ozurumba hacks Ahunze to death and sets her house on fire, trapping the village in a state of fear. He also objectifies her through the excision of her body parts. Rather than creating a safer community, shaming the "violators" of social norms potentially leads to crimes that are more egregious.

The novel's exposition of the target's experience of shame is enabled through a third-person omniscient narrator with absolute access to the character's interiority. With such access to the thought process behind Ozurumba's murder-suicide, *I Saw the Sky Catch Fire* suggests that genital shaming may strengthen the offender's transgressive behavior, problematizing the putative beneficial nature of shaming. In this context, perhaps the performers of the cursing or shaming ritual should consider the standing of the individual(s) within the community and whether they would risk communal disapproval and taking into account that the punitive act may fail against several groups of individuals: those suffering from mental illness; those having no social position to protect, and the fear of losing that will be cause for conformity; those lacking obligations to the normative codes; and those who have severed their social and communal ties and discarded their social identity as husband, nephew, or father. If the gesture does not work with these lowly groups of men and does not work with powerful men, two groups that are most likely to trample communal values, who *does* the cursing work on? This question continues to haunt the necessity of these cursing and shaming gestures.

In the novel, the agency of women, whether Nne-nne or the market women, is temporary and often volatile because their targets reacted with murderous violence. However, like the women's agency, the colonial administration's ideological power and agency are also temporary because decades after the event, the scholarly intervention of female anthropologists, political scientists, and sociologists uncovers women's contribution to struggles against colonization. That said, showcasing women's agency has often fallen into the easy trap of triumphalism. For instance, the absence of the soldiers in Ugbala's call, symptomatic of dualistic thinking (the women versus the colonial administrator instead of the triadic women, soldiers, and colonialists), has also resulted in misreadings of the novel.

Teresa Washington's study uses the Igbo indigenous system of signification, and more specifically the figure of Aje, to read the scene of women's public exposure as uncritically empowering: "Although condemned to eternal victimhood by some feminists and international organizations, Africana women possess abundant revolutionary skills and abilities, and they have used them against slavers, colonizers, and their oppressors" (2005, 134). To uphold/promote an indigenous mode of being, Washington's approach constructs women as endlessly powerful, thereby complicating commonplace images of women as victimized by their traditions. Consequently, her analysis is more a response to unflattering images of African women in the media than a close reading of the novel. Similarly, in her insightful discussion of translation in Echewa's novel, Bella Brodzki highlights "female ingenuity, courage, autonomy, and survival in the face of male impotence and unreliability" (2007, 210). Yet this interpretation flattens out the range of possibilities intrinsic to genital cursing/shaming, including its unexpected and counterproductive effects. In retrieving and "brandishing" authentic African women's forms of power, literary critics often misrepresent the experiences of female characters, resulting in sweeping generalizations.

These well-meaning postcolonial analyses not only overlook the kinds of violence (epistemological as well as physical) that genital cursing/shaming begets but also ignore the evanescence of positions of power, thus reifying the dichotomy between triumph and victimhood that often informs much scholarship on the ritual display of bodies. More broadly, these studies play on an unquestioned understanding of relations of power as if decades of poststructuralist theoretical developments, which reached their peak with Michel Foucault (1997) and Judith Butler (2016), did not problematize conventional understandings of power as an established domination that emanates from a specific space. One thus detects in this uncritical approach what Lila Abu-Lughod (1990) has called "the romance of resistance" in her study of the Bedouin of Egypt, or what Amanda Anderson (2000) calls "aggrandized agency"—the tendency to rescue uncritically from oblivion hitherto ignored forms of contestation. Recent backlash against defiant disrobing invites us to be attentive to the unstable nature of positions of power and victimhood. The urgency of the task is in keeping with the proliferation of naked protest in Africa and the increasing availability of more fictional narratives.

If we agree with Jonathan Culler that "the novel serves as the model by which society conceives of itself, the discourse through which it articulates the world" (1975, 189), it becomes apparent that *I Saw the Sky Catch Fire* is more nuanced in its position on insurgent disrobing than are individual stories of emancipation and subjugation. These "single stories," as Chimamanda Ngozi Adichie (2009) calls them, currently permeate scholarship on naked protest in the African context and internationally. The reactions and the cultural and social identities of the women's targets impact, often significantly, the force of women's intentional nudity. For instance, the restaging of the murderous feedback of the soldiers, the revenge of Nne-nne's relatives, and the murder-suicide of Ozurumba suggest that the women's targets have a higher level of agency, the deployment of which reduces women's agency. By acknowledging both the potentialities and limits of genital-based cursing and shaming practices, literary criticism and fiction may become a starting point for a more effective symbology of political action.

Defiant Disrobing Going Viral

News reporting about protesters using exposed breasts and genitals as forms of contestation has been proliferating around the world. Naked agency has profusely crossed borders because of globalization and its attendant wider circulation of images and ideas. This border crossing—because of borrowing, imitation, or the proliferation of exceptional circumstances—suggests two conclusions: that the oppositional tactic of one group can serve as a model for other resistance groups, and that the performance of insurgent self-exposure in one place can dilute or enhance its reception elsewhere. Specifically, can the proliferation of naked protest, or the ubiquity of sexual images in daily lives, dramatically influence its potency in Africa? Or in America?

"Going viral" is a phrase that captures the speed and ease with which social activists and other protesters borrow or influence the use of defiant disrobing from one another. The social capital of sexuality, the media attention that it guarantees, and, most important, the availability of nakedness and feelings all contribute to its virality. Donna Sheehan's candid report of her borrowing of naked exposure, from the July 2002 Nigerian women's Chevron Texaco protest, crystallizes this dynamic. Sheehan organized her November 2002 nudity protest in California, where it became one of the first and most publicized female collective deployments of uncivil nakedness in the United States. She attributes the genesis of her idea of women's disrobing to oppose then-president George W. Bush's international policies and the United States' possible military intervention in the Middle East to the 2002 Nigerian women's example.

It was a Sunday morning when I recalled the dream upon waking. I dreamt of people creating artistic shapes with their bodies. My thoughts went to Helen Odeworitse and 600 Nigerian women, who used the threat of their nakedness (a shaming gesture for men in Nigeria and perhaps elsewhere?) to force Chevron Texaco to listen to their families' needs. That was such a powerful image for me at the time that it became a natural extension of my thoughts about my dream. Now I saw women's bodies forming letters—and the word they formed had to be PEACE. (Sheehan 2002b)

Borrowing or imitating does not, however, guarantee that the original meanings will travel unscathed. At the heart of borrowing is the possibility, or even the guarantee, that the borrowed form will be adapted differently in its new context. For instance, although the rhetoric of peace and vulnerability is central to Sheehan's formulation, that is not the message that journalists and scholars have attributed to the Niger Delta women, or what the African women told themselves and others after reaching the deal with multinational oil companies. According to one of their leaders, Helen Odeworitse, their goal was to inflict pain in the form of shame and curse on the male oil company executives. This meaning of weapon, not just symbolically but also materially, differs slightly from the meaning that Sheehan assigns to her practice: "Baring Witness is about using the greatest weapon women have, the power of the feminine, the power of our beauty and nakedness to awaken our male leaders and stop them in their tracks. . . . By risking with our nakedness—our charm and beauty and vulnerability—in service of peace we are exposing the flesh all humans share. We are casting off the old dominant paradigm of aggression and restoring the power of the feminine to its rightful place as the protector of life" (2002a).

If Sheehan's uncivil self-exposure moves away from violence and destruction for the restoration of feminine power, for the Niger Delta women, the destructive is also constitutive. The latter's act does not need a politics of civility, which is paramount within the dominant North Atlantic form of sociality and politics. Étienne Balibar has criticized Western obsession with civility in his topography of contemporary violence, *Violence and Civility* (2015). In the schema that Balibar interrogates, politics is regarded as civility, which ceases where violence begins and vice versa. Given that violence is tragically inherent within the practice of politics, Balibar argues, politics can never abandon violence, especially in the form of resistance.

Despite these different meanings and significance, a few months after Sheehan's protest, several news sources reported that thirty women lay naked in the snow in New York's Central Park in February 2003 to protest against the war in Iraq (*Sunday Mail* 2003). Although no independent source cited the California protest as the model directly copied by protesters in New York City, Paul Reffell, a member of Sheehan's organization (Baring Witness), establishes a causal link between them, stating, "The photograph of their protest became the shot seen around the world, once it hit the news wires and the Internet. It has aroused passion and inspired women and men nationwide to take action, speak their minds and express their frustrations at not being heard by those in power" (2002). Reporting on the defiant self-exposure by a group of Cape Town women, Tricia Shannon (2003) goes even further than Reffell to argue that the vision and practice of a female bodily based model of resistance link Niger Delta (Nigeria) to California and New York City (United States) to Byron Bay (Australia) to Cape Town (South Africa). Following the causal link, it seems that this nonlinear yet possible transmission brings American, Nigerian, South African, and Australian protesting women into a globalized world. They are united around pain and body, despite the historical, ideological differences inscribed on women's bodies. These protesters, separated by thousands of miles yet connected through hope and digital media, suggest that the female body, independently of age, race, location, and socioeconomic status, is a possible site of contestation and self-empowerment.[1]

Inspiration, however, has failed to translate into equal access to the conventional channels of knowledge production. Observers and scholars have acknowledged the contribution of Nigerian women to other activist movements, but there exists an imbalance in the quality of the archive about the Nigerians. Although social scientists and others have access to American and European activists' thoughts and strategies through the activists' websites and books, these scholars lack similar access to the practical and conceptual strategies that Nigerian and other African women have mobilized. Consequently, most contemporary studies of naked exposure quote American and European activists at length, referring to their books and websites and reducing the complexity of Nigerian women's acts or threats thereof to mottoes and footnotes. For instance, Helen Odeworitse's protest is often mentioned with her declaration, "Our nakedness is our weapon," which still stands as a cultural curiosity (BBC News 2002).

The dearth of material on the feelings, strategies, and rationale of the Nigerian women increases the possibility of the women being ventriloquized. Put differently, despite the wider circulation of their threat of nakedness, the archive on their rationale is sketchy or inaccessible to a global audience. Reasons for this relative absence remain elusive and diverse. Historical inequalities may determine who has access to online media and the conventional channels of knowledge production. Or, as I have argued with several examples, the Nigerian women's refusal to interact with journalists may have been because of taboo and variables of age and gender. To be fair, perhaps the Nigerian women had no interest in sharing their rationales and thoughts. Alternatively, perhaps the difficulty of accurate translation from their local languages to global languages such as English or French is the major obstacle. Independently of the reasons, it is undeniable that although an overflowing archive exists about female defiant nakedness in the African context, the archive remains overwhelmingly about, and not authored by, women participants.

Given the absence of the women's voice, two trends on the meanings of these defiant self-exposures, as well as their relation to capitalist "modernity," emerge. Typical reactions explain women's threat to use their tabooed body parts by assigning priority either to sociocultural factors that relate to modernization or to the essence of African societies themselves. Whereas some take African cultural practices out of their historical and political contexts, others consider the practices as the reemergence of the forces of backwardness. Still others (elected local officials, the women themselves, academics, and non-African artists and filmmakers) use the gesture and women's images for waging ideological and political battles or for creating economic self-empowerment. In that sense, there is little difference between these positions with their unitary readings; all offer an etiolated account of the dynamic of power relations that defiant disrobing ignites. I have moved beyond the congratulatory appraisal and the condemnatory rhetoric in order to examine defiant disrobing as both an expression of women's resistant agency and an index of the changed political landscape in Africa.

In addition to the rhetoric of backwardness and nativism is the idea of peaceful protest dilutes the insurgence effects of the women's act with the idea of peaceful protest. Popular responses in online news sources, comments on social media, and academic treatises have overwhelmingly interpreted defiant disrobing in ways that are at odds with the few intended meanings that the protesters themselves revealed. For instance, journalists or observers present these defiant disrobings as peaceful forms of action à la Gandhi or Martin Luther King Jr., whereas actual participants hope to highlight their violence by causing multiple misfortunes for the targeted males and/or for society.

Despite the possible borrowings, the specific meanings of naked protest have not traveled widely, seemingly going viral either almost stripped bare or overdressed in meanings. It is in this sense that the "social life," to use Arjun Appadurai's (1986) concept, of retaliatory nakedness is vibrant yet precarious. In a single instance of public disrobing, multiple meanings may cohabit: for example, the belief that mature women's bodies house powerful entities that may cause the social or biological death of their targets, and the women's refusal or inability to speak the language of liberal democracy. In Britain, the effects of nakedness evoke questions of hygiene and propriety, and the disruption of the quotidian. In the British 1986 Public Order Act, the naked person in public, whether protesting or not, may be charged with "disorderly, threatening, abusive, or insulting behavior, likely to cause alarm, harassment and distress."[2]

As I have demonstrated throughout this book, the meanings of intentional nakedness are markedly contextual, and culturally and historically determined. Yet mobilization of nakedness in one space may affect its reception in another because of the interconnected nature of our world. Such deployments include the celebratory performance of bodies in antiglobalization, feminist, and queer activism in the dominant Euro-American context with social movements including Occupy Wall Street, lactivism, World Naked Bike Riders, and PETA's Running of the Nudes protest against bullfighting; as well as with the playfulness of fads such as streaking, mooning, and flashing.[3] Put differently, how does a resident of Palm Beach, Florida, accurately assess the serious circumstances that led the women of Manipur, India (2004), or the farmers of Veracruz, Mexico (2002–8), to expose themselves against gendered violence and corporate land grabs? In the fast-paced world of news production and consumption, little time is available for news reporters and viewers to deliberate on defiant nakedness enacted by either dispossessed farmers in Veracruz or naked bike riders in San Francisco.

The coexistence, and even the collision, of meanings of uncivil nakedness remains a documented fact, and various meanings need not be transcended for a universally singular and unquestionably accurate meaning. However, this coexistence and collision call for reflection if the reactions, predicated upon local discourses of sexuality and protest, turn a protester's most desperate act into the ludic activism that Occupy Wall Street protests and anti-Trumpists have promulgated. Further deliberation is necessary if the ultimate weapon for some women becomes merely one more playful spectacle of sexualization and globalization. Self-exposure as playfulness may create indifference in the viewer, or provoke only amusement, thus obscuring the dangers to which some protesters in highly oppressive and volatile contexts are exposed. The differential exposure of protesters to violence becomes clear with the understanding of the direct correlation between the proliferation of protests and the inadequacy of institutional channels of liberal democracy either to address issues or to hear the voices of discontent. It is in exposing these entanglements that *Naked Agency* contributes significantly to a deeper understanding of embodied modes of political action in Africa and around the world.

Introduction

1 Alaimo 2010; Barcan 2004; Carr-Gomm 2010; Lunceford 2012; Souweine 2005; and Stevens 2006 establish a direct link between these events.

2 However, when relevant, I use the term *genital cursing* to signal what the gesture is called in the published literature.

3 Three versions of *Nudes and Protest* are available online. I analyze the green-and-blue version available in Judeh 2012. Despite its seminal status as one of the first visual renderings of defiant self-exposure in Africa, Bruce Onobrakpeya's plastographic artwork has received relatively little scholarly attention. See the second and third versions in Mwantok 2016, and MutualArt's *Nude and Protest* (Red Line Experiment) 2007.

4 *Biopolitics*, the political rationality undergirding the government of the polis, is used here rather than *biopower* because of the former's encompassing capacity. Biopolitics includes biopower, the simple capacity to legislate sovereignty and the strategies dictated by biopolitics. Another conceptual distinction is grounded in Foucault's deeper definition of resistance as not just negation but also, and perhaps more importantly, as creation (1978, 1979, 1980, 1982, 1997). According to Maurizio Lazzarato's "From Biopower to Biopolitics" (2002), the move from biopower to biopolitics reflects Foucault's theoretical development. Given these distinctions, I deploy biopolitics in my engagement with other theorists of power, including Agamben, Achille Mbembe, and Michael Hardt and Antonio Negri, for whom biopower is a less generative concept. For more on the distinction, see also Lemke 2011.

5 For more on the implications of "naked life," see de la Durantaye 2009; Robert 2013; Salzani 2012, 2016.

6 Discussing the ubiquity of nakedness in Trump's road to the White House, Barbara Browning (2016) highlights the accessibility of the gesture to all, making it perhaps one of the most available strategies, one that is assumed to be almost universally accessible.

7 The term *woman* does not indicate my reification of *women*. The fact that only a certain category of females could feel the power to inflict injury on the body politic underscores how *woman* is not a single category in these societies (Amadiume 1987; Hoffman 1998; Oyěwùmí 1997).

8 *Africa* here refers to sub-Saharan Africa and its multiple forms of governance, cultural practices, and historical conditions. *Africa* should not be understood as a homoge-

neous continent; I take seriously Mudimbé's (1988) reflections on the fictionality of the continent.

9 Several Euro-American documentaries have featured the playfulness of protest disrobing, e.g., the French *La nudité toute nue* (Nicklaus 2007) and *La face cachée des fesses* (Pochon and Rothschild 2009), and the Canadian *Naked* (Bissell 2000). *Naked* is not distributed but I obtained a copy from the filmmaker.

10 "Il y'a une sorte de force de la nudité aujourd'hui dans la pensée, et sans doute je crois parce que nous sommes dans un temps qui se sent particulièrement exposé, et particulièrement dévêtue de vêtements idéologiques" [A strong sense of nudity pervades our thinking. That is so, I think, because our era is one that feels particularly exposed, that is particularly stripped of ideological vestments]. All translations are mine unless otherwise noted.

11 See, for example, Amadiume 2003; Ardener 1973; Awasom 2003; Bastian 2002; Brett-Smith 1995; Buckley and Gottlieb 1988; Conrad 1999; Diduk 1989; Drewal and Drewal 1983; Ekine 2008; Grillo 2018; Hodgson 2017; Jackson 2012; Kah 2011; Nkwi 1985; Prince 1961; Ritzenthaler 1960; Serubiri 2017; Shanklin 1990; Stevens 2006; Tanga 2006; Tripp et al. 2009; Van Allen 1972; Wipper 1982.

12 The use of anthropological texts to contextualize the protest and its effects is fraught with tension, since the texts cannot be taken at face value. In fact, anthropological research should be understood in the larger and contentious conversation around cultural anthropology and its objects. In *The Invention of Africa: Gnosis, Philosophy, and the Order of Things* (1988), Valentin Yves Mudimbé discusses the role that anthropology has played in reifying Africans and their cultures as abnormally different from an imagined Western-universal norm. For more on the controversy over anthropology, see Jane Guyer's *Marginal Gains: Monetary Transactions in Atlantic Africa* (2004), in which she echoes Mudimbé: "'The cultural particularism of anthropology' . . . tends 'to make Africa look like a pathological departure from a standard model based on Western experience and institutions'" (172). These reflections serve as a cautionary measure against either investing the value of science in anthropological descriptions of the rituals or rejecting them. Since these texts are nearly the only option for making contact with these cultural practices, their outright rejection will not only curtail the possibilities of understanding these societies but also constitute an epistemic violence against these rituals and the women who draw strength from them.

13 For references to *bottom power*, see Baker and Mangan 1987; Day 2008; Hodgson and McCurdy 1996; Matory 1993; Stephanie Newell 2002; Ogunseitan 2010; Ogunyemi 1996, 2007; Okeke-Ihejirika 2004; Viola, Bardolph, and Coussy 1998.

14 If seniority was crucial to the effectiveness of genital cursing among the Igbo, it was otherwise elsewhere, especially among the Beti (Ombolo 1990). According to Jean-Pierre Ombolo, women of all ages had the capacity to curse targeted males.

15 Kassim Koné, email communication, March 2013.

16 Similar paradoxical understandings of female genitals as both destructive and life giving have been observed throughout the world. In *A Brief History of Nakedness* (2010), Philip Carr-Gomm lists the uses of naked female genitals in ancient rituals in South American, Indian, and European folk magic, in which they were designed to

attract love and to enhance the fertility of the land. For example, women would walk, dance, and sing naked and sometimes urinate in the fields, or under the moonlight, "to encourage the growth of crops, or assist in finding a love partner or drawing closer the love interest." In that case, "the women were using their nakedness in another form of fertility magic to encourage the evolutionary aim of partnership: the birth of children" (Carr-Gomm 2010, 38). Carr-Gomm not only highlights the influence of these practices on rituals of modern witchcraft but also links them to contemporary events, such as the urgent attempt of fifty Nepali women in 2006 to induce rainfall by ploughing naked in their drought-ridden fields. One of the ritual participants was quoted in a local newspaper as saying, "This is our last weapon, we used it, and there was light rainfall" (Carr-Gomm 2010, 40). In "Nudity" (1987), Arvind Sharma provides a list of auspicious meanings bestowed on the female naked body, ranging from rain-making and crop success to human fertility (8–9). The ancient Greek story of Baubo, who cheered up the grieving goddess Demeter by shaving off her pubic hair and exposing her genitals to her, is another frequently cited myth about the therapeutic properties of female self-exposure (Chausidis 2012; King 1986; Marcovich 1986; Olender 1985). Similar accounts of the apotropaic nature of female genitalia have been reported in India, Japan, China, and Ireland. For authoritative texts on these, see Dexter and Mair 2004; Freitag 2004; Guest 1937; Pearson 1997 and Rhoades 2010 on Ireland; Donaldson 1975; Schmidt 1987 and Sonawane 1988 on India; Le Anh 2004 on China and Japan; and Witzel 2005 on Japan and India.

17 "Par ailleurs, le sexe, surtout de la femme, est tenu chez les Africains pour une réalité extrêmement dangereuse: capable de produire la vie, il est suprêmement capable de la supprimer."

18 Kassim Koné, email communication, March 2013.

19 Although Misty Bastian's studies (2001, 2002, 2005) are about the Igbo-speaking peoples of Nigeria, her findings resonate with those of other scholars, including Sarah Brett-Smith about the Bamana (1995), Shirley Ardener (1973) and Robert Ritzenthaler (1960) about the Kom, and Jean-Pierre Ombolo about the Beti (1990). Around the world, scholars have uncovered the potential cosmological destructive power of female bodies. In *Encyclopedia of Esoteric Man* (1977), Benjamin Walker reports, "The vagina is a destructive orifice and a devirilizing element in the female structure, and sexual intercourse results in a kind of castration. Woman then becomes the devouring female, the vulva incarnate" (305). This myth is restaged in cultural productions such as Lichtenstein's comedy/horror movie *Teeth* (2008). In *Nudity: A Cultural Anatomy* (2004), Ruth Barcan also notes the image of the *vagina dentata*, toothed vagina, and the association of nudity with witches and witchcraft. Barcan highlights the contradictory and rich meanings associated with female nakedness and contrasts ancient times with contemporary. In ancient times, female nudity, like the phallus, was vested with magic power. However, Barcan claims that the long history of wild, powerful, sinister, or sacred meanings of female nakedness are being displaced by the modernization process, with its attendant "curbing and limiting of some of the archaic meanings of female nakedness and the domestication of its wilder meanings" (191).

20 In *La guerra de las gordas*, translated into English as *The War of the Fatties and Other*

Stories from Aztec History (1994), Salvador Novo dramatizes how during the war with the Mexicas the battalion of armed men of the Aztec army was defeated by naked women and their milk: "The most hair-raising battalion bursts into our ranks. They came up shouting and slapping themselves on the belly. We were so shocked we couldn't move. And when they got close to us, they squeezed their chichis and bathed our faces with squirts of warm, thick milk!" "The secret weapon! The atomizing pump!" (54).

21 A similar conceptualization of openness is Charles Taylor's "open space" (2017) wherein it stands at the midpoint of multiple prevailing perspectives and benefits from their vantage points. Additionally, Carole Boyce Davies's influential concept of the migrating subject (1994) has paved the way for naked agency.

22 The edited volume *Vulnerability in Resistance* makes a similar point. See, for instance, Judith Butler's "Rethinking Vulnerability and Resistance" (2016).

23 OED Online, s.v. "Naked, adj.," Oxford University Press, accessed January 8, 2019, www.oed.com.

24 OED Online, s.v. "Nude, adj.," Oxford University Press, accessed January 8, 2019, www .oed.com.

25 These distinctions have been the subject of many studies. Compounding the divide, art historian John Berger draws on work in the aesthetic vocabulary to distinguish between nudity and nakedness in relation to the act of seeing and being seen, declaring, "To be naked is to be without disguise. The nude is condemned to never being naked. Nudity is a form of dress" (1972, 54). Linda Nead (2002) and Rob Cover (2003) reject the binary for different reasons. Whereas Nead highlights the gendered aspect of the binary, Cover considers poststructuralism and its emphasis on the loose boundaries between sexuality and obscenity, which he calls "derailed categories." Although cogent, Cover's attempt is a refusal to acknowledge the prevalence of systems of hierarchy that still structure the postmodern world.

26 Despite claims about the secular nature of the West and despite the cultural diversity therein, it is impossible to think about nakedness outside the Christian tradition (Agamben 2011).

Scene 1: Exceptional Conditions and Darker Shades of Biopolitics

1 For an account of the incomplete nature of Foucault's reflections on biopolitics, see Thomas Lemke's *Biopolitics: An Advanced Introduction* (2011). Foucault's accounts of the concepts of biopower and biopolitics are articulated in *The History of Sexuality*, vol. 1, *The Will to Knowledge* ([1978] 1998); *Security, Territory, Population: Lectures at the Collège de France, 1977–1978* (lectures of January 11, and February 8, 15, and 22) ([2004] 2009); and "Lecture 11 of 17 March 1976," in *"Society Must Be Defended": Lectures at the Collège de France, 1975–1976* ([1997] 2003).

2 There is a twist in this Swiftian Modest Proposal if Summers's memo was serious. Jonathan Swift's *A Modest Proposal* (1729) is a satire in which he jokes about eating Irish babies.

3 For more on the controversy about this alleged memorandum, see Enwegbara 2001; *Harvard Magazine* 2001; Mittelman 2009; Swaney 1994.

4 For a theoretical discussion of the issue, see Hoad 2005.

5 In three lectures (4, 5, and 11 in January, February, and March 1976) of *"Society Must Be Defended": Lectures at the Collège de France, 1975–1976* ([1997] 2003), Foucault mentions colonization fleetingly, attributing to it the incipience of "racial racism," which by way of the "boomerang effect" is reworked in Europe. For instance, in "Lecture 11 of March 17, 1976," Foucault says, "Racism first develops with colonization, in other words, with colonizing genocide. If you are functioning in the biopower mode, how can you justify the need to kill people, to kill populations, and to kill civilizations? By using the themes of evolutionism, by appealing to a racism" (257). Despite Foucault's scant attention to Africa, colonization contributes productively to his work on biopolitics, which he attributes to colonial racism through a boomerang effect, the importation of colonial models into the West. Martinican poet, philosopher, and statesman Aimé Césaire, in *Discours sur le colonialisme* (1955), made a similar argument more than two decades before Foucault's lectures, as did Hannah Arendt in *Origins of Totalitarianism* ([1948] 1973).

6 However, the court has been scathingly critiqued by scholars of Africa (Grovogui 2015; Villa-Vicencio 2000) and public officials as "the invisible hand of neocolonialism," my play on Adam Smith's metaphor the "invisible hand" ([1759] 1781).

7 Esposito proposes his thesis through a series of rhetorical questions: "Is this how biopolitical discourse ends? Is the only possible outcome of such events such an overlapping, or is there another way of practicing, or at least thinking, biopolitics, which is to say a biopolitics that is ultimately affirmative, productive, and removed from death's nonstop presence? In other words, is a politics no longer over life [*sulla vita*] but of life [*della vita*] imaginable? If it is, then how should it, how might it, take shape?" (2013, 77).

8 In early reflections, Esposito used the term *the immunization paradigm* to designate the process whereby the individual continually attempts to gain immunity from the suffocating, and yet nurturing, effects of the community, to which she initially surrendered power for protection (Esposito and Campbell, 2006, 28).

9 The centrality of the hydraulic metaphor of flow in these accounts of globalization has been problematized with the rocky and uneven metaphor of "hops." For more accounts, see James Ferguson (2005) in his analysis of resource extraction in sub-Saharan Africa and Anna Tsing's (2004) carving of global channels in her work on deforestation in Indonesia.

10 In "What Is a Destituent Power?" (2014), Agamben takes as his point of departure constituent power that he finds to reproduce governmental structures of exception and as the pinnacle of metaphysics. He then proposes destituent power as an antidote to constituent power.

Scene 2: Dobsonville and the Question of Autonomy

1 Teresa Barnes reports, "The film clearly angered many black women (and men). . . . They questioned how the video had been made, by whom it was seen, and how the community of squatter women had been involved in the product" (2002, 252).

2 These documentaries highlight the forms of contestation, notably public disrobing and genital cursing, that emerge in response to the widespread sense of perishability and injustice prevalent in postcolonial nation-states. Exploitative economic measures and the environmental degradation caused by multinational oil companies in the Niger Delta (Schermerhorn 2010), as well as the protracted, brutal civil war in Liberia (Disney 2008), led to the women's defiant disrobing or threat thereof. For more on this, see "The Cinematic Language of Naked Protest" (Diabate 2007).

3 The procedure is named variously depending on different ideological positions. The neutral terms include *female genital cutting* (FGC), *female genital circumcision* (FGC), *female circumcision*, and *female genital surgeries*. The most controversial term is *female genital mutilation* (FGM). For more on this, see Hodžić (2017) and Nnaemeka (2005). In this book, I will use terms according to their relevant contexts.

4 On female genital circumcision, see Arnfred (2004), Gunning (1998), and Oyěwùmí (2003).

5 Writing of the Niger Delta women's showdown with multinational oil companies, Stevens (2006) and Turner and Brownhill (2004) reach similar conclusions.

6 Section 10 of the apartheid-era Urban Areas Act grants urban housing rights to black male South Africans who were born in or had lived in an urban area for more than fifteen years. But "the right to housing was confined to male heads of household. Women were defined as minors under customary law and thus dependents of men. So single women, widows and divorcees could not access housing in their own right, as every woman was defined as subject to a man. The effect of these policies was to exclude single women from the right to rent their own homes" (Meintjes 2007, 350).

7 Grossberg separates agency from subjectivity and the subject: "Agency—the ability to make history as it were—is not intrinsic either to subjectivity or to subjects. It is not an ontological principle that distinguishes humans from other sorts of beings. Agency is defined by the articulations of subject positions into specific places (sites of investment) and spaces (fields of activity) on socially constructed territorialities. Agency is the empowerment enabled at particular sites and along particular vectors" (1993, 15).

8 Scholars have indicated the female-gendered nature of spirit possession in most societies in Africa where it has been observed. Anthropologist Paul Stoller (1995) identifies five dominant explanatory models of spirit possession: functionalist, psychoanalytic, physiological, symbolic (interpretive), and theatrical. For more on spirit possession in Africa, see Bastide 1978; Boddy 1989, 1994; Mudimbé 1988; Rouch 1957.

9 On the goals and processes of conversion of the native to Christianity, Valentin Mudimbé writes, "A person whose ideas and mission come from and are sustained by God is rightly entitled to the use of all possible means, even violence, to achieve his objectives. Consequently 'African conversion' came to be the sole position the African could take in order to survive as a human being" (1985, 154). On a similar

point, Achille Mbembe emphasizes that conversion "involves an act of destruction and violence against an earlier state of affairs . . . [that] is always carried out in the name of a specific materiality, one that claims to oppose a system of truth to an order of error and falsehood" (2001, 230). The vast literature on conversion tackles questions such as "reverse conversion" and conversion as a two-way experience, although the missionary–indigenous encounter was rigged by the fundamental inequalities of the colonial context. Specifically, Comaroff and Comaroff, in their analysis of the nineteenth-century London Missionary Society mission among the southern Tswana peoples of South Africa, emphasize, "The missionary encounter must be regarded as a two-sided historical process; as a dialectic that takes into account the social and cultural endowments of, and the consequences for, *all* the actors—missionaries no less than Africans" (1991, 12). See also Gerbner 2015 and Stanley 2003.

10 In *South African Yearbook 2001/02* (Republic of South Africa: Government Communications), the 2001 census numbers show that almost 80 percent of South Africa's population identified as members of a Christian denomination. The 2011 census did not ask about religion.

11 Recently, Annie Coombes (2001, 2003), Pumla Gobodo-Madikizela (2004), and others have analyzed the forms that South African women's militancy have taken since the implementation of apartheid.

12 In "Agency and Pain: An Exploration" (2000), Talal Asad reminds us that in modern "rational" societies, agency is attributed to individuals so they can be held accountable in a court of law, which turns agency into "the product of pain" (34).

13 African philosophy offers an array of studies on the gap between the dominant Western liberal conception of the person as a self-determining autonomous individual and the conception in Africa of the self as open-ended and ever-evolving. See Kwasi Wiredu and Kwame Gyeke's germane collection of essays, *Person and Community: Ghanaian Philosophical Studies* (1992), and D. A. Masolo's *Self and Community in a Changing World* (2010).

14 To choose Deleuze and Guattari is to already have taken a position in the debate around the so-called inadequacy of Western theories to explicate texts and cultural practices occurring in the postcolonial African context. As I see it, the use of Eurocentric theories as reading lenses counters notions of nativism and indigenism while favoring a more integrative human community. To be more specific, I qualify my use of Deleuze and Guattari as a critical cosmopolitanism or cosmopolitan comparativism, which Arnold Krupat (2002) has argued for in the Native American studies context. Like Krupat, I see the value and meaning of the indigenist, the nationalist, and the Eurocentric approaches, and as a comparatist, I strive to link them all without any claims to transcendence. Kenneth Harrow (2002) cogently discusses the question.

15 For more on the 1929 and 1959 beer hall protests, see la Hause 1982; Minkley 1996; Sadler 2002; C. Walker [1982] 1991.

Scene 3: Africanizing Nakedness as (Self-)instrumentalization

1 "L'Adjanou, la danse sacrée et rituelle des femmes en pays baoulé qui avait aidé Houphouët, au temps colonial, à prendre la victoire sur les colons, est sortie pour purifier, chasser, exorciser et conjurer tous les mauvais sorts."

2 "A l'appel des jeunes patriotes de manifester contre l'attaque qui a coupé la Côte d'Ivoire en deux le 19 septembre 2002, j'ai demandé aux danseuses d'Adjanou de mon village de s'organiser pour conjurer le malheur. Nous avons dansé pendant sept jours. A la fin de cette semaine mystique, les rebelles ont débarqué dans le village, et on emporté les danseuses qui ne sont plus jamais revenues."

3 See also the works of Simplice Allard (2011), Macaire Etty (2010), Salyff G. (2011), Zéré de Mahi (2010), K. Kouassi Maurice (2011), and many others who penned opinion papers in 2010–11, reminding Ivorians of the massacre of the priestesses in order to shape their electoral decisions or to oppose Ouattara.

4 In this 1997 study of political power, Séhoué proposes to uncover the close connections between the presidential palace, the president, and Adjanou dancers. The author tells the story of Djaa Kouakou, an albino who was bequeathed in 1979 at age sixteen to the Boigny family. Kouakou's mystical power as an albino was channeled to heal the president, whom modern medicine and Swiss doctors had failed to cure. Mamie Faitai, the president's sister, is described as the "Minister of Occult Sciences" in charge of all that is connected to "fetishisme, de l'organisation des sacrifices, de la prise de contact avec les marabouts, les oracles ou d'autres formes de spiritualité traditionnelle" [fetishism, the organization of sacrifices, contacts with Marabouts, oracles, and other figures of traditional spirituality] (Séhoué 1997, 57). The preface of the book, written by Don Mello Ahoua, a prominent member of the opposition party, suggests its biased nature. The more occultist the ruling party appears, the more rational the opposition frames itself. By highlighting occultism as the center of the ruling regime, the opposition seeks to show its inability to lead the people into the twenty-first century. This book should therefore be understood within the broader political context of the country, especially with the 1990s reintroduction of political pluralism.

5 Laura Grillo (2018) uncovers similar bodily adornments among the Abidji of Côte d'Ivoire.

6 "Ne doit pas être pris à la légère, car on ne s'amuse pas avec les esprits et on ne ruse pas avec Dieu."

7 "De toutes les manifestations feminines, celle qui était la plus chargée de sens, c'est l'adjanou. Celle qui est la plus importante, la plus grave c'est l'adjanou."

8 "J'entendis resoner mon adjanou. Je mis mon 'kodjo' et nous dansâmes l'adjanou, faisant le tour entier de la ville pour porter chance à nos hommes, pour aider nos hommes à réussir. Puisque nous n'avions pas de fusils, nous nous sommes dit, il faut que nous soyons nombreuses pour que les Blancs sachent que Houphouët est soutenu."

9 "La victoire était petite mais combien belle et qui méritait d'être exploitée."

10 "A nos vaillantes femmes qui, par leur marche historique sur la prison de Grand-Bassam, le 24-12-1949, ont arraché la liberté confisquée des hommes / 06-02-1999 / Maire F. Ablé."

11 The 16th Summit of the French president and more than thirty African heads of state

was held on June 20, 1990, in La Baule (a French department). In his speech, Mitterrand promised debt relief to African countries that would implement democracy.

12 My discussion of the Usana and their publicized contribution to the Casamance
conflict resolutions builds on Vincent Foucher's (2006, 2007) and Beniamina
Lico's (2014) fieldwork and publications. For more on the Usana and similar female
religious-mystical groups, see Maria Teixeira (2001).

13 For more on Track II diplomacy, see Diamond and McDonald 1996; Jones 2015. Conceived in the 1970s, Track II diplomacy became formalized in the 1990s. It designates
the practice of non-state actors in peace-making processes, turning them into active
agents in their communities and countries rather than mere recipients of a top-down
approach.

14 In Senegal and elsewhere around the world, sacred groves play significant indigenous
religious roles within communities that keep them. They serve as sites of religious
rites, festivals, and recreation. Paradoxically, in Senegal, urbanization and the
emotional and material challenges of westernization may have increased the need for
sacred groves in urbanized centers, where groups of individuals would lay claim to a
forest fragment as their sacred grove.

15 For example, Carol Gilligan (1982) and Sara Ruddick ([1989] 1995) analyze the specific forms of doing politics that women bring to the public sphere. These include "the
ethic of care and maternal thinking": caring, mothering, and peacefulness (Gilligan
1982). In the African context, such feminist work has taken the form of showing the
defining contribution of women's roles as cooks, caregivers, and cleaners to the success
of indigenous movements of liberation (Alam 2007).

16 In *Njinga of Angola: Africa's Warrior Queen*, Linda Heywood (2017) argues that
despite the comparable contribution of Njinga of Angola to Queen Elizabeth I of
England and Catherine the Great of Russia, her monograph, *Njinga of Angola,* is the
only serious biography of the African queen.

Scene 4: In the Name of National Interest

1 The Liberian Ministry of Foreign Affairs issued a statement claiming that Ghanaian authorities "brought about an end to a month-long protest by Liberian refugees
in Buduburam Camp, thereby facilitating UNHCR access to the Camp to resume
normal humanitarian work, including provision of food to the refugees that had been
interrupted for a month" (Government of the Republic of Liberia 2008).

2 The government invoked a clause of the 1951 Refugee Convention relating to the
status of refugees to force the UNHCR to close operations for more than twenty-five
thousand Liberian refugees in Ghana. The clause stipulates that when conditions
have improved in a refugee's country of origin, the host government is no longer
obliged to host them. For more on the convention, see the UNHCR's "Protocol and
Convention Relating to the Status of Refugees" (2010).

3 The dictator novel is a literary genre that originated in Latin America. Although it
dates back to the nineteenth century, the genre flourished in the 1960s and 1970s to
challenge the power of authoritarian figures. The dictator novel often deploys formal

elements such as fragmentation, stream of consciousness, interior monologues, multiple narrative points of view, frequent lack of causality, neologisms, and unconventional narrative strategies, which resonate with the grotesque. For more, see Marjorie Agosin (1990). For an analysis of *Wizard of the Crow* as a dictator novel, see Robert Spencer (2012).

4 A graduate with a BA in economics and an MBA earned in India (paid after arduous financial sacrifices by his parents), Kamiti wa Karimiri returns home to a series of unsuccessful job interviews and bouts of hunger that lead him to forage mountains of garbage. After three years in this predicament, an unsuccessful and humiliating job interview at Tajirika's Eldares Modern Construction and Real Estate Company, where he meets Nyawira, Tajirika's secretary, defeats him.

5 A quick keyword search for *shame* in Heinemann's African Writers Series database (1962–2003) of about 250 texts yields the following results: 170 novels, 57 poetry entries, and 15 drama entries. Of 242 entries with more than 1,010 hits, almost 99 percent include the word *shame*. Among texts with the term *shame*, some are canonical and some are newer. These include Chinua Achebe's *Things Fall Apart* (1958), Ama Ata Aidoo's *Changes* (1991), Kofi Awoonor's *This Earth, My Brother* (1972), Mariama Bâ's *Une si longue lettre* (1981), Mongo Beti's *Remember Ruben* (1974), Calixthe Beyala's *Tu t'appelleras Tanga* (1988), J. M. Coetzee's *Disgrace* (1999), Ahmadou Kourouma's *Les soleils des indépendances* (1968), Kopano Matlwa's *Period Pain* (2016), Thando Mgqolozana's *A Man Who Is Not a Man* (2009), Ousmane Sembène's *Guelwaar* (1996), and Sony Labou Tansi's *The Shameful State* (2016). African Writers Series, accessed December 18, 2017, http://collections.chadwyck.com/marketing/home_aws.jsp.

6 This expansive conceptualization of shame is uncovered in multiple colonial and postcolonial African contexts, notably those of the Malinké, the Hausa, and the Yoruba, where female defiant disrobing has been observed. For more information, see Babatunde 2000; Keita 2011; Kirk-Greene 1974; Lawuyi 2012; and Vuarin 2000.

7 That menstruation carries stigma in most contexts is undeniable, although Erica Johnson and Patricia Moran's (2013) establishment of a direct correlation between femininity and shame does not necessarily pan out in the African contexts under consideration here. To avoid setting up the generalized African context as antithetical to the dominant West, I recognize that most postcolonial contexts have adopted Judeo-Christian precepts of decorum that underpin political secularism.

8 To these assaults on their dignity and freedom, women have fought back to reclaim a space of agency. For more on their resistant protests, see associated Press 2014.

9 In the novel, three garbage collectors discover a "lifeless" body in a pile of garbage. They search the dead man for money and valuables but find nothing but rags. The collectors curse the corpse as if it were alive, saying, "You stupid liar. I am sure these rags are your real clothes and the suit you have on is stolen property. *Have you no shame*, stealing other people's clothes? And you did not even have the good sense to steal a suit less worn out; at least we could have taken that" (40, emphasis mine). This episode also allows the narrator to account for the pervasiveness of deaths in the novel, some of which are due to hunger, illness, or suicide.

10 In scene 8, I take up Echewa's work for a different set of questions. For that reason, I will focus on Ngũgĩ's text in this scene.

11 In *Women of the Forest* (1974), Yolanda Murphy and Robert Murphy report a similar cartography of modesty among the Munduruku Indians of Brazil.

12 Jean-Pierre Ombolo stresses the importance of disrobing in rituals of initiation, healing, and purification in precolonial Beti society. According to him, a woman suffering from sterility will undergo a public performance of a ritual of healing in which she stands in the middle of the audience, which repeats incantations after the officiant. Then, the naked person is smeared with the blood of a sacrificed animal, in this way transferring the evil from the sick person to the animal (1990, 91). Alternatively, in the case of incest, the culprits will appear naked in public for expiation. In the case of adultery, to dishonor the woman, she is stripped naked in public and excised or circumcised (Even 1939; Ombolo 1990). Even in fight situations, stripping the other naked is the unmistakable sign of victory. Ruth Barcan notes the ongoing power of nakedness in contemporary consumer societies, like those of Europe and the United States, in the humiliating power of the stripping of victims in torture such as the Abu Ghraib case and the importance of stripping naked in "initiation rites and bastardization practices" (2004, 134).

13 "On croit que la nudité est exigée uniquement, parce que si l'on a des vêtements sur soi au moment de la cérémonie, on est considéré comme lié avec eux. . . . Mais les conjurés veulent aussi et surtout signifier par leur nudité, la droiture du coeur, la bonne foi qui règnera dans leurs rapports avec les frères."

14 According to Misty Bastian, "men supply cloth for their wives' 'attires'; women choose cloth for their husbands' outfits or buy them ready-to-wear shirts and trousers in the marketplace; mothers procure cloth and sew outfits for their children, sometimes cutting up their own attire to make that worn by their offspring; sisters lend lengths of expensive wrapper cloth to their brothers for important events; patrons give their clients cloth with their own names, symbols, or likenesses printed on it as a form of publicity for their alliance. Wealthier neighbors share their old clothes with poorer neighbors or with more distant relatives. People are known abroad by their clothing, and that clothing is scrutinized for signs of its social provenance" (2005, 38).

15 No historical documents confirm Eisenhower's visit to Rwanda. The nature of Estés's story (in parts fictional, historical, and analytical) shows that the distinction between fiction and history is problematic.

16 Here I use the term *performance* in the double sense of the public display of technical skills by an actor and acting for purposes of mobilization for sociopolitical change. I follow Marvin Carlson's two distinct concepts of performance (1996). Elaborating on Richard Schechner's (1973) analysis, Carlson privileges a combined conceptualization of the "act" that seeks recognition of a skill, and the "act" in protest against social and political injustice, noting that "performance is always performance for someone, some audience that recognizes and validates it as performance even when, as is occasionally the case, that audience is the self" (1996, 5).

Scene 5: Film as Instrumental and Interpretive Lens

1 For more on these films, see Diabate 2017.

2 The second prize included a check of 5 million CFA francs, which is approximately $11,000.

3 The other three rhetorical questions are "How can you make an action film in a country where acting is subversive?," "How can you make a crime film in a country where investigation is forbidden?," and "How can you film a love story where love is impossible?"

4 "Lors d'un séjour au Cameroun, je voulais rencontrer un ministre mais n'y parvenais pas. Une jeune fille m'a dit qu'elle pouvait organiser ça. J'ai vite compris que c'était tout un réseau et que j'aurais pu rencontrer tout le gouvernement ainsi! Ces jeunes filles maîtrisent le fonctionnement du pays et ont un certain pouvoir."

5 "Quant aux filles, elles sont un sujet intéressant car lorsqu'elles ont un comportement ambivalent, tout le monde s'en mêle: les traditions, la police. . . . C'est moins radical lorsqu'il s'agit de garçons. Elles me permettent de politiser mon regard sur la société."

6 Now an established filmmaker on the West African cinema scene, Bekolo achieved continental and international recognition in 1992 at the age of twenty-six thanks to *Quartier Mozart*. Bekolo studied screen editing at the Institut national de l'audiovisuel (INA, National Audiovisual Institute) in France in 1988–89 and later worked for Cameroon Radio and TV (CRTV), editing short films and video clips for African musical groups. He also directed several documentaries and a television series, and taught filmmaking and African cinema in several American universities. Despite these experiences, his creativity was tested and established with the limited budget of $300,000 with which he shot *Quartier Mozart*. To overcome the challenges of distributing the film in Douala, Cameroon, a city with only two main commercial theaters, the director screened it in the city's town hall for weeks and advertised it on taxicabs.

7 The choice of cinema as a means of both documenting the litany of problems afflicting Africa and suggesting solutions is compelling because the medium was part and parcel of the apparatuses of colonization. Regardless of its early contribution to the colonizing mission, cinema today constitutes the primary medium of reaching the largest number of West Africans. Cinema is popular in Africa for several interrelated reasons. First, most of the population is nonliterate, and cinema allows nonliterate people a space of self-representation that increasingly closes the social divide between themselves and the literate. Second, Africans in general have learned to distrust the written document, which is closely associated with colonialism and its repressive mechanisms. Lastly, the cinematic text's ability to provide a more authentic representation of socioeconomic realities makes it an increasingly appropriate medium for political and social intervention (Niang 1996; Syrotinski 2002).

Although these arguments have been useful in understanding the popularity of cinema in Africa and its importance in political exigencies, they may have overlooked the evolving material realities and the powerful impact of modernizing forces. For one thing, the preeminence of the visual is not exclusive to the African continent. Second, the rapid expansion of mass communications (radio, television, and newer information technologies, such as the internet), the continued growth of filmmaking techniques, and the sprawling market economy that allows most households access to a television set are some of the contributing factors. Various other influences, local and foreign, foster enthusiasm for cinema in black Africa. Continuing to dwell on Africans' supposed inclination toward orality would be to subscribe to imaginary ontological differences.

8 Although the messages of Bekolo's productions have been considered inaccessible to mass audiences, Bekolo still envisions his films as entertaining as well as educational. On his professional blog, he has written that he sees filmmaking and public service as two sides of the same coin, adding, "It's all about education. The tools are there, the interest is there, we just need to create a method of acquiring knowledge that uses what is today the most immediately accessible medium" (2006). As the director uses the blogosphere to bridge the gap between his productions and his audiences, in *Les Saignantes*, he uses the science fiction genre to attract a larger young African audience.

9 "J'ai découvert le Mevoungou à travers un roman, *Le Tombeau du soleil* de Philippe Laburthe-Tolra (Le Seuil/Points Odile Jacob) qui enseigne à la Sorbonne" (quoted in Barlet 2005).

10 "Le Mevungu en revanche, se présente clairement, du moins pour ses adeptes, comme une célébration du clitoris et de la puissance féminine."

11 "Les femmes vont l'admirer (le clitoris) et se frotter contre lui . . . , on va enfin le chatouiller, le masser ou l'étirer; 'jusqu'à l'amener à la longueur d'un membre viril.'"

12 Most anthropological reports consistently document the importance of women's bodies in restoring social balance and prosperity to the community, but they differ on other uses that may appear subversive to patriarchal values. At times, the Mevoungou ritual is described as women's mode of resistance against male domination. The ritual functioned, among other uses, as a means of "affirming [women's] personality, of reinforcing their productivity, and of engaging in a double sexuality," that is, sanctioned bisexuality (Bochet de Thé 1985, 248). However, most male anthropologists and intellectuals, French or Beti, dismiss those subversive goals for which the initiates may have used the ritual. The debate surrounding the ritual concerns its cultural value, its sexual components, and its significance. For instance, Charles Gueboguo (2002) claims that the descriptions of the ritual suggest that it is a celebration of women's sexual organs and power and admits that it does contain homosexual gestures. However, he opposes its eroticization, arguing that its most important use was for procreation (Gueboguo 2002, 35). For Laburthe-Tolra, the ritual was not a means of resistance but a celebration of women's sexual organs. He suggests that it resembles a desperate plea to summon the powers of nature to heal the community rather than the constitution of a female world of pleasures (1985, 331–32).

Cameroonian intellectuals, including Catholic priests Paul Mviena (1970) and Isadore Tabi (1991), as well as Bekolo himself, echo Laburthe-Tolra and Gueboguo in commenting on the sexualized interpretations that they think French feminist anthropologists bring to bear on the ritual. Mviena argues that the ignorance of its real significance and the misinterpretations by colonialists justified its abolition. The Catholic priest pleads that the Mevoungou not be regarded as "un Sabbath des lesbiennes" (a Sabbath of lesbians) because it was exclusively performed for procreation and social purposes (Mviena 1970, 162). In 2005, more than thirty years after Mviena, Bekolo subscribed to the desexualization of the ritual, claiming that it focuses on a force of nature, women's sexual organs, to heal the community and did not implicate sexual relations: "En beti, la même racine désigne le concubinage, la descendance, le procréateur etc.: au moins une dizaine de termes sans rapport avec l'acte sexuel" [In Beti, the same root (Mevoungou) refers to cohabitation, offspring, progenitor, etc.: at least fifteen words with no relation to the sexual] (quoted in Barlet 2005).

13 Known in Côte d'Ivoire as *la position du tir*, the "shooting position" is a dance move created in 2002 by young Ivorian artists imitating the shooting positions of soldiers and rebels during the 2001 armed rebellion that divided the country for more than a decade.

14 Postcolonial African bureaucrats, in their majority, condone such rejection (Fisiy 1998). Aspiring to "modern" forms of being, ruling classes set modernity against rituals that they construct as markers of primitive societies. Ironically, though, while they publicly disavow occultist practices, they resort to them for their personal gain, as does the minister of state in the film. It is therefore subversive in the eyes of the ruling classes that *Les Saignantes* advocates an intelligent use of occult forces such as the Mevoungou ritual as an antidote to the failures of the postcolonial state.

15 For representations of this trope in African films, see Sembène Ousmane's *Xala* (1975), Souleymane Cisse's *Yeelen* (1987), Benoit Lamy and Ngangura Mweze's *La vie est belle* (Life is rosy) (1987), and Bekolo's *Quartier Mozart* (1992).

16 *Demoiselles de Porto-Novo* (2012), a play on Pablo Picasso's *Les Demoiselles d'Avignon* (1907), is an intriguing series of photographs by Leonce Raphael Agbodjélou that stages black girls with naked torsos and wearing sculpted masks.

17 Such need falls loosely in the category of internet addiction. For more on this, see Cash, Rae, Steel, and Winkler 2012; Davidow 2012.

Scene 6: Secularizing Genital Cursing and Rhetorical Backlash

1 Controversy around the ritual led the French newspaper *Courrier International* (2003) to print a translated version of the story. The original report (September 28, 2001) is in English with the title "Gambia: In Election Ritual Saga Naked Women Protest in Brikama."

2 Here, my methodological approach is to gather material through proxy interview. Although the Jolas of The Gambia share many similarities with the Dioulas of Côte d'Ivoire (my ethnic group), significant differences exist between them. Jallow, a fellow comparatist, agreed to do the interview with the Jola women on my behalf because I do not speak the Gambian Jola language. Siga Jallow, email communication, June 13, 2014.

3 It is the framing of women's genitalia as "intimate" that undergirds the title of Laura Grillo's powerful reflection on instances of "female genital power" in *An Intimate Rebuke* (2018).

4 The dictatorial regime of Yayah Jammeh showed no interest in arresting the women.

Perhaps the fact that the opposition was the target of the women's wrath worked in favor of the ruling regime, which would naturally refuse to undermine its chances at reelection by arresting those who curse the opposition.

5   John Mbiti (1969) considers Christianity and Islam as African religions.

6   Outtara (2012): "ces pratiques dignes de l'Afrique 'sauvage' et 'brutale' n'ont pas leur place dans les pays civilizés."

7   Abidjan.net is the major web portal in Côte d'Ivoire. It brings information from all Ivorian newspapers and allows readers' comments, a functionality that newspapers' websites do not offer. Controversy about the necessity of readers' comments arose with Pam Belluck's "Comment Ban Sets Off Debate" (2013) and Dominique Brossard and Dietram A. Scheufele's "This Story Stinks" (2013). In the African context, several edited collections have engaged the rising trend of readers' comments. See Okoth F. Mudhai, Wisdom Tettey, and Fackson Banda's *African Media and the Public Digital Sphere* (2009), Karin Barber's *Readings in African Popular Culture* (1997), Sokari Ekine's sms *Uprising: Mobile Activism in Africa* (2010), and Stephanie Newell and Onookome Okome's *Popular Culture in Africa: The Episteme of the Everyday* (2013).

8   Originally located in the south and central west of Côte d'Ivoire, the Bété people are one of the Ivorian major ethnic groups. For more on the history and customs of the Bétés, see Dozon 1985a, 1985b.

9   Kouassi Koffi, Kouassi Kouakou, Dosso Farouk Stephane, Fofana Vaz, Mamadou Karamoko, Lewest Payspropre, Adama Koné, Moustapha L'areigné Bamba, Taragone Dekasséré, Dan Dany, and Marie-Chantale.

10   "Ces vulgaires 'djosseuses' et 'mon mari m'a chassé,' sans aucune pudeur, pour la plus part, sont nées pendant la 'veillée funèbre' de la honte. Elles donnent l'occasion aux Français, de distinguer, entre la diaspora ivoirienne de la france, la bonne graie [*sic*] feminile [*sic*], qui a une dignité pour leurs corps et respect des valeurs republicaines françaises, de l'ivraie, que ces 'djantras' démodées representent [*sic*]."

11   "C'est une autre forme de sauvagerie que de se mettre en de tel habillement dans une contrée qui n'est pas la sienna."

12   The 1958 anlu uprising in the Bamenda Grassfields in British Cameroon against local men and the British colonial administration caught the attention of several scholars (Ardener 1973; Awasom 2003; Diduk 1989; Ritzenthaler 1960).

13   A similar account of the futility of papers to legitimize land ownership occurred in northern Tanzania among the Maasai. A village government member complained to Benjamin Gardner: "They [Thomson Safaris] are owning the land through papers. That land is our traditional land. They have the right and protection of the government if they believe they own the land through papers. . . . Who to believe, paper or us? They have a legal paper from Dar es Salaam. What about the people in this area? This is our ancestors' land" (2016, 119). In his monograph *Selling the Serengeti: The Cultural Politics of Safari Tourism* (2016), Gardner reports that the Maasai's lament echoes that of many: that the written laws and contracts were used by the government to dispossess them from their land rather than to protect their rights to their land as citizens (146).

14   "Oui, notre Nation s'écroule et nous devons prier pour que Dieu daigne la relever." By

appealing to what appears to be a Christian "God," the women demonstrate that their confidence in the potency of their indigenous mystical forces appears shaky.

15 "Vraiment on est dépassé par tous ces évènements, on en peut plus. . . . Gbagbo a trop humilié les Baoulés, je crois que cette fois-ci c'est son terminus à Yamoussoukro avec les femmes."

16 "Devant donc cette réaction, nous resterons nues devant la cour jusqu'à ce qu'on sache pour quoi cela a été fait. . . . Venir nous voir nues, nous ne pouvons pas accepter cela."

17 In April 2016, the organization Activate posted to YouTube a forty-five-minute theater production titled #RUReferenceList: Welcome to the Zoo. The play features female and male and black and white performers in white tops and dark pants. ("Welcome to the Zoo," April 27, 2016, accessed July 12, 2018, https://www.youtube.com/watch?v=_GmllFkj6p8&t=896s).

scene 7: Epistemic ignorance and menstrual rags in paris

1 Drawing on the nationalist rhetoric of the women's rallies, I use the generalizing and descriptive category of "the French political class" that the self-described cursing Ivorian women oppose in their rallies. Nationalism was central to the women's perceived takeover of the country's autonomy. For instance, during the rallies, rather than incorporate newly created songs that publicly expose the targets' offenses to further compound the necessity of the curse, the women relied on popular music to communicate their patriotism. Their theme song was the popular "David contre Goliath" of the Ivorian musical group Espoir 2000. This song borrows from the biblical story of David versus Goliath (the Ivorian people represented here by David and France by Goliath) and features the refrain, "Yé i yé éh, l'ennemi ôh, petit à petit s'est dévoilé" (Yé i yé éh, little by little the enemy has uncovered himself). This song castigates the intervention of the French army in Côte d'Ivoire while celebrating Gbagbo's sense of patriotism and his noble goal of preserving the dignity of black people. However, the long-standing entanglement of Côte d'Ivoire and France and the presence of Ivorian-French in the French administration, as well as the migration of thousands if not of millions of Ivorians to France, belie the dichotomous opposition between French and Ivorians in France.

2 Organized in France, the negotiations were facilitated by representatives from France, the United Nations, the African Union, and the Economic Community of West African States (ECOWAS). The warring Ivorian parties were Le Front Populaire Ivoirien (FPI), Le Mouvement des Forces d'Avenir (MFA), Le Mouvement pour La Justice et La Paix (MJP), Le Mouvement Patriotique de Côte d'Ivoire (MPCI), Le Mouvement Populaire Ivoirien du Grand Ouest (MPIGO), Le Parti Démocratique de Côte d'Ivoire (PDCI), Le Parti Ivorien du Travail (PIT), Le Rassemblement Démocratique Africain (RDA), Le Rassemblement des Républicains (RDR), L'Union Démocratique et Citoyenne (UDCY), and L' Union pour la Démocratie et la Paix en Côte d'Ivoire (UDPCI). For more on the negotiations, see United States Institute of Peace, "Côte d'Ivoire: Linas-Marcoussis Agreement" (2003). For more on the crisis, see Banégas and Losch (2002).

3 For more on the stalled peace agreement, see N. Cook 2011; Piccolino 2014.

4 Other women's movements participated in the show of strength, including the one led by Claudine Ouattara, who traveled to France after de Villepin's visit to submit to the Quay d'Orsay her organization's plan for the peace negotiations in Côte d'Ivoire. In an infantile move, however, Ouattara told journalists: "Nous lançons un cri du coeur à la France. Nous nous sentons comme un enfant abandonné par son protecteur. On veut voir à nos côtés 'nos ancêtres les Gaulois'!" [We're issuing a distress call to France. We feel like a child abandoned by her protector. We would like to see on our side "Our Ancestors, the Gauls"!] (quoted in Hofnung 2003).

5 "Un mode de protestation que craignent tous les présidents africains: leurs agents de sécurité ne peuvent pas regarder Les femmes nues qui symbolisent leurs mères et urinent pour chasser les mauvais esprits."

6 "[Mr. de Villepin] en sait quelque chose depuis son voyage mémorable en Côte d'Ivoire en tant que ministre du gouvernement français." ["(Mr. de Villepin) can tell you one or two things about his memorable trip to Côte d'Ivoire when he was minister of domestic affairs"].

7 The inclusive name L'Association des Femmes Patriotes Ivoiriennes de France (AFPIF) appears as a smoke screen despite the fact that the women's protest demonstrations for the former Ivorian president Laurent Gbagbo weaken the call for inclusivity. Beyond Paris, the mobilization of the Kodjo Rouge has mushroomed in Côte d'Ivoire during perceived moments of political and social duress. One of these moments concerned the illegal dumping of toxic waste in Abidjan that significantly endangered the health of local populations. After a minister embezzled the fines paid by the shipper and related entities to compensate victimized populations, these populations organized their kodjos rouges to "curse" their opponents while drawing attention to their cause (Aka 2012; Dassé 2016).

8 "Culotte sans attache entre les fesses et les cuisses," "loincloth."

9 For more on the systematic inclusion of indigenous Ivorian language words in French, see Gandonou 2002; G. Kouassi 2007; Lafage 2003. Cursing demands words because of the way in which they challenge the obviousness of the distinction between subject and object and awaken latent contradictions in the Europeans' own relationships to material things.

10 See literary fiction such as Jean-Marie Adiaffi's *La carte d'identité* (1980), *Une vie hypothéquée* (1984), *Silence, on développe* (1992), and *Les naufragés de l'intelligence* (2000); and François-Joseph Amon d'Aby's *Kwao Adjoba* (1956).

11 "Le kodjo a sa source dans la pure tradition africaine. Il est un morceau de pagne de couleur principalement rouge servant de cache-sexe à la femme avant que la modernité occidentale avec son slip, n'entre dans les moeurs de la femme africaine. Comme cache-sexe, le kodjo entre dans la symbolique de la pudeur féminine mais n'est pas que cette symbolique. Il symbolise aussi le pouvoir de maudire de la femme, celle qui donne vie, et dont la parole est sacrée, quand elle l'exhibe ou l'ôte pour proférer des malédictions."

12 For more on this formulation, see Hutchison 2012.

13 "Un pouvoir incommensurable de maudire ceux qui en sont les auteurs et les responsables" (Togui 2012c).

14 "Cette cérémonie est l'occasion [pour nous] d'agir pour la chute de Nicolas Sarkozy. La malédiction par le 'Kodjo' tombera sur Nicolas Sarkozy et il perdra les prochaines éléctions."

15 Cursing is manifestly the ability of spoken words to create reality, not just the everyday reality. It is a more powerful reality, as Adame Ba Konaré argues in her historico-anthropological study of words among the Malinké. In *L'os de la parole: Cosmologie du Pouvoir: Essai* (The bone of the word: Cosmology of power) (2000), she provides the following Malinké/Bamanan saying: "Toute chose engendre son enfant et que seule la parole engendre sa mère. En d'autres termes, la parole peut avoir des consequences majeures, facheuses et inattendues, au point de perdre son propre auteur" [All living entities engender their progenitors, only words birth their mothers. Stated differently, words can have massive, unfortunate, and unexpected consequences to the point of endangering or harming their speaker] (11).

16 "Celui qui n'est pas sorti du sexe d'une femme pour venir à la vie, seul celui-là, peut échapper à la colère de la femme dans cette circonstance.'" This statement was made during the rally and reported by blogger Louis-Freddy Aguisso on May 5.

17 "Elles ripostent par le moyen qui leur sert de bouée de sauvetage, en tant que dépositaire de la vie; parce qu'elles sont les mères des hommes."

18 In these analyses, spirits and gods appear without being consigned to the prepolitical, as Guha observes in the Indian case in "The Prose of Counter-insurgency" ([1983] 1994); see also Chakrabarty's "Minority Histories, Subaltern Pasts" (1998).

19 "C'est pourquoi, lorsque la femme se trouve devant une injustice criarde irrésistible, elle recourt à ce qui dans son corps mais aussi dans son esprit, fait sa force. Et ses atouts se résument à son sexe. En Afrique, la femme ne montre sa nudité à tout le monde que dans des circonstances très graves telles que le deuil ou pour appeler la malédiction sur quelqu'un ou sur quelque chose, mais jamais pour faire de la publicité."

20 Numerous operations of the U.S. Army abroad include Operation Enduring Freedom, Operation Relentless Strike, Operation Desert Shield, Operation Desert Storm, and Operation Iraqi Freedom. Scholars from multiple disciplines have noted the shift in how these names are now coded: from previously dry and uninspiring to rhetorically powerful. For more on the shift, see the October 2009 special issue of the *PMLA*, "War" (Aravamudan and Taylor); Ibekwe and A. Adebayo 2012; Jaeger, Baraban, and Muller 2012; Joint Chiefs of Staff 1993.

21 "Mais quand nous observons, nous voyons que les gens ne veulent pas nous écouter et maintiennent toujours les gens en prison. Ils ont toujours gardé notre président Laurent Gbagbo en prison. Blé Goudé est parti le rejoindre à la CPI."

22 These include le Courant de Pensée et d'Action de Côte-d'Ivoire (COPACI, School of Thought and of Action in Côte-d'Ivoire), Le Congrès pour la Résistance et la Démocratie (CRD, Congress for Resistance and Democracy), Le Comité de Défence des Institutions et de la Souveraineté de la Côte d'Ivoire (CODESCI, Committee for the Defense of Ivorian Institutions and Sovereignty), and the movement of "3000 Women for Côte d'Ivoire" (Togui 2012).

23 The coalition united the CODESCI (Committee for the Defense of Ivorian Institu-

tions and Sovereignty), the Cri panafricain (the Panafrican Rage), and the Collectif des Français d'origine ivoirienne (CFIACI) (Collective of French of Ivorian Origin) (Togui 2012e).

24 In her research on the Bamana of Mali, anthropologist Sarah Brett-Smith (1995) reports from the village elders the appropriate ways to respond to genital cursing.

25 For more on this question, see Chatin 2016; Fonte 2008; Fullilove 2008; Kagan 2009; Kelly 2007; Plattner 2008; Slaughter 2004.

26 "[Elle] fera sans doute date dans l'histoire de la résistance pacifique des patriotes ivo-iriens de la diaspora contre le néo-colonialisme et le despotisme en Côte d'Ivoire. Ce fut le rendez-vous de l'indignation et de la révolte. Mais sans couteaux, sans kalach-nikov, sans bombe. Et pourtant ces femmes déterminées ont mené l'offensive contre Alassane Dramane Ouattara et Barack Obama. Pour cette énième offensive, elles ont sorti l'arme fatale, le Kodjo rouge Terminator."

27 "Les coeurs des ivoiriennes saignent, suite aux nombreux morts causées par les bom-bardements de l'armée française de la résidence du président Gbagbo et des ivoiriens, sous les ordres de Nicolas Sarkozi [sic]."

28 Scholars of social movements have shown the importance that songs and dances play in democratic processes on the continent. Songs inspire solidarity, communicate movement beliefs, express emotions, and allow participants to partake of collective identity. The women's theme song, "David vs. Goliath," is part of the repertoire of songs and dance moves that opponents mobilized in reaction against the oppressive regime of Houphouët Boigny in the 1980s–1990s and against the 2002 rebellion. See Boka 2013; Drewett and Cloonan 2006; Foua 2011; Sasha Newell 2012; Schumann 2010.

29 Article 222-32 of the penal code: "L'exhibition sexuelle imposée à la vue d'autrui dans un lieu accessible aux regards du public." Erica Ho (2012) reports how the city tries to ensure public decency. And that means to enforce a few rules that may seem foreign to many French beachgoers. The police specify, "Any outfit that allows for the genital area or breasts to be seen constitutes sexual exhibition and is punishable by a year in prison" (Ho 2012).

30 Several scholars have noted the challenge that social movements and political parties face in maintaining the proper balance between signaling numerical strength while adhering to a radical sociopolitical agenda (Simmel [1908] 1950; Wouters and Van Camp, 2017).

scene 8: (Mis)reading murderous reactions

1 In an interview with Anwesha Das, Echewa claimed: "I believe that my identity as an Igbo, Nigerian, and African became crystallized when I crossed the border into an-other country and hemisphere. In that sense, living in the United States has made me more consciously African than I would have been, had I remained in Nigeria all these years" (Das 2014, 158). He has authored several literary texts, including *The Land's Lord* (1976), *The Crippled Dancer* (1986), *The Ancestor Tree* (1994), and *The Magic Tree: A Folktale from Nigeria* (1999). For more on Echewa, see Das 2014.

2   The first recorded public exhibition of bodies against British colonialists occurred on March 14, 1922, during the Harry Thuku protest in Kenya. The British arrested the supposed menace Harry Thuku, an esteemed nationalist with a pan-Africanist vision who had encouraged the natives, especially women, to boycott the British-imposed programs. During an attempt by an estimated eight thousand people to secure his immediate release, Mary Muthoni Nyanjiru made history by leaping to her feet and exposing her body with these words of defiance: "You take my dress and give me your trousers. You men are cowards. What are you waiting for? Our leader is in there. Let's get him" (Rosberg and Nottingham [1966] 1985, 52). Her passionate invitation was followed by a scuffle, during which Nyanjiru and 20 others died, according to the highly disputed colonial archives of deaths. Some gave the number as between 25 and 250 (Alam 2007; Kanogo 1987). Eventually, Thuku was released and exiled to northern Kenya from 1922 to 1930.

3   Ironically, the colonial administration's eventual imposition of a poll tax on women was a double-edged sword because the primary decision to remove them from taxation was an expression of sexism. By reading Igbo society through the lenses of British culture, the administration misread women's economic independence and ignored that, either individually or collectively, women provided men with resources to pay the taxes, thus avoiding the incarceration of men and husbands. Of course, the taxation of women did not mean that the colonial officials were feminist or necessarily in favor of female freedom. Rather, it was an indication of the imperial project's dire need for resources.

4   Following the war, the colonial empire put together the Aba Commission of Inquiry to shed light on the disturbances. In 1930 the commission published *Notes of Evidence Taken at the Commission of Inquiry Appointed to Inquire into the Disturbances in the Calabar and Owerri Provinces, December 1929.*

5   Judith Van Allen (1976) argues that whereas "women's war" stressed the prominent role of women and the idea of a grave assault on an enemy, the British localized, diminished, and diffused the impact of the revolt. Nigerian scholar Chikwenye Ogunyemi further complicates the naming of the Women's War with her coding of the event as the "women's struggle" (1996, 53). She thinks that "war" refers to a neat, closed, and time-situated category, whereas "women's struggle," an ongoing effort, more accurately reflects women's resistance.

6   Using Charles Kingsley Meek's (1937) reports is a challenging methodological decision because he was a British government anthropologist sent to investigate the origins of the war. However, his and others' texts remain the most accessible sources on the war.

7   The social power of obscenity in the songs can best be understood in the context in which deference and respect toward men and husbands are the central components in the structuration of social relations (Ardener 1973).

8   In contradistinction to received ideas, the prevalence of the theory of sex pollution of the female body is prevalent in the context of a relatively balanced power dynamic between men and women. Indeed, as Mary Douglas has suggested in her seminal text *Purity and Danger*: "When male dominance is accepted as a central principle of social

organisation and applied without inhibition and with full rights of physical coercion, beliefs in sex pollution are not likely to be highly developed. On the other hand, when the principle of male dominance is applied to the ordering of social life but is contradicted by other principles such as that of female independence, or the inherent right of women as the weaker sex to be more protected from violence than men, then sex pollution is likely to flourish" ([1966] 2003, 176).

Epilogue

1 Globalization has played a crucial role in inspiring a culture of contestation around the world. The most recent case is what is now problematically referred to as the "Arab Spring," a site of contentious campaigns, performances, and displays that generated ripple effects across Europe (Dabashi 2012; Howard and Hussain 2013).

2 According to Carr-Gomm, "Being naked in public is not necessarily unlawful in Britain. It is an offense if the police believe and can prove that the act intends to offend someone. It is framed more in terms of intentionality than the state of undress" (2010, 167).

3 Over centuries in Europe, some of these practices held major resistant meanings that have been eroded with industrialization (Barcan 2004).

Aba Commission of Inquiry. 1930. *Notes of Evidence Taken at the Commission of Inquiry Appointed to Inquire into the Disturbances in the Calabar and Owerri Provinces, December 1929*. London: Waterlow.

Abidjantalk. 2013. "Togo: Les vieilles femmes nues font fuir les soldats de gnass." January 16.

Ableman, Paul. 1984. *The Banished Body*. London: Sphere.

Abu-Lughod, Lila. 1990. "The Romance of Resistance: Tracing Transformations of Power through Bedouin Women." *American Ethnologist* 17, no. 1 (February): 41–55.

Achebe, Chinua. 1958. *Things Fall Apart*. London: Heinemann.

Achebe, Chinua. 1970. "The Duty and Involvement of the African Writer." In *The Africa Reader: Independent Africa*, edited by Wilfred Cartey and Martin Kilson, 168–69. New York: Random House.

Achebe, Nwando. 2005. *Farmers, Traders, Warriors, and Kings: Female Power and Authority in Northern Igboland, 1900–1960*. Portsmouth, NH: Heinemann.

Adélé, Alexis. 2016. "Côte d'Ivoire: Ulcérés par les 'microbes,' les habitants d'Abidjan se font justice." *Le Monde*, April 4.

Adesokan, Akin. 2008. "The Challenges of Aesthetic Populism: An Interview with Jean-Pierre Bekolo." *Postcolonial Text* 4, no. 1: 1–11.

Adiaffi, Jean-Marie. 1980. *La carte d'identité*. Paris: Hatier.

Adiaffi, Jean-Marie. 1983. *The Identity Card*. Translated by Brigitte Katiyo. Harare: Zimbabwe Publishing House.

Adiaffi, Jean-Marie. 1984. *Une vie hypothéquée*. Abidjan: Les Nouvelles Editions Africaines.

Adiaffi, Jean-Marie. 1992. *Silence, on développe*. Paris: Nouvelles du Sud.

Adiaffi, Jean-Marie. 2000. *Les naufragés de l'intelligence*. Abidjan: CEDA.

Adichie, Chimamanda Ngozi. 2009. *The Danger of a Single Story*. TED Talk. https://www.ted.com/talks/chimamanda_adichie_the_danger_of_a_single_story?language=en. Accessed July 12, 2013.

Afigbo, Adiele E. 1982. "The Native Revenue Ordinance in the Eastern Provinces: The Adventure of a Colonial Legislative Measure." In *Studies in Southern Nigerian History*, edited by Boniface Obichere, 73–102. London: Frank Cass.

AFP. 2011. "Manifestation pro-Gbagbo à Paris." *Le Figaro*, April 10.

AFP. 2012. "Malgré l'opération Kodjo rouge Terminator: La CPI rejette la demande de mise en liberté provisoire de Laurent Gbagbo." AFP, October 26.

Agamben, Giorgio. 1991. *Language and Death: The Place of Negativity*. Translated by Karen Pinkus with Michael Hardt. Minneapolis: University of Minnesota Press.

Agamben, Giorgio. 1995. *Homo sacer: Il potere sovrano e la nuda vita*. Vol. 2. Turin: G. Einaudi.

Agamben, Giorgio. 1998. *Homo Sacer: The Sovereign Power and Bare Life*. Translated by Daniel Heller-Roazen. Stanford, CA: Stanford University Press.

Agamben, Giorgio. 2000. *Means without End: Notes on Politics*. Translated by Vincenzo Binetti and Cesare Casarino. Minneapolis: University of Minnesota Press.

Agamben, Giorgio. 2005. *State of Exception*. Translated by Kevin Attell. Chicago: University of Chicago Press.

Agamben, Giorgio. 2011. "Nudity." In *Nudities*, translated by David Kishik and Stefan Pedatella, 55–91. Stanford, CA: Stanford University Press.

Agamben, Giorgio. 2014. "What Is a Destituent Power?" Translated by Stephanie Wakefield. *Environment and Planning: Society and Space* 32, no. 1 (February): 65–74.

Agbodjélou, Leonce Raphael. 2012. *Demoiselles de Porto-Novo*. Satchi Gallery, London. Photographs, C-print. https://www.saatchigallery.com/artists/leonce_raphael _agbodjelou.htm.

Agosin, Marjorie. 1990. "Inhabitants of Decayed Palaces: The Dictator in the Latin American Novel." Translated by Barbara E. Pierce. *Human Rights Quarterly* 12, no. 2: 328–35.

Aguisso, Louis-Freddy. 2012. "France: L'effet Kodjo rouge en marche." *La Côte d'Ivoire Debout*, May 5.

Agustín, Laura María. 2010. *Sex at the Margins: Migration, Labour Markets and the Rescue Industry*. London: Zed Books.

Ahmadu, Fuambai. 2017. "Equality, not Special Protection: Multiculturalism, Feminism, and Female Circumcision in Western Liberal Democracies." In *Universalism without Uniformity: Explorations in Mind and Culture*, edited by Julia L. Cassaniti and Usha Menon, 214–38. Chicago: University of Chicago Press.

Aidoo, Ama Ata. 1993. *Changes*. London: Women's Press.

Ajayi, Femi. 2010. "The Appalling Approach of Nigerian Women Involvement in the 2011 Election." *Nigeria World*, September 3. http://nigeriaworld.com/columnist /ajayi/090310.html.

Ajayi, Kunle. 2010. "Exploring Alternative Approaches for Managing Electoral Injustice in Africa: The Case of Breast Protests in Nigeria and the Sex Strike in Kenya." *ISS Paper* 214 (August): 1–19. https://www.files.ethz.ch/isn/120979/214.pdf.

Aka, Marcelle. 2012. "Déchets toxiques/Des victimes initient les opérations 'Kodjo rouge' et 'Casserole.'" *L'Inter*, May 29.

Akyeampong, Emmanuel. 1996. *Drink, Power and Cultural Change: A Social History of Alcohol in Ghana, c. 1800 to Recent Times*. Oxford: James Currey.

Alaimo, Stacy. 2010. "The Naked Word: The Trans-corporeal Ethics of the Protesting Body." *Women and Performance: A Journal of Feminist Theory* 20, no. 1 (March): 15–36.

Alam, S. M. Shamsul. 2007. *Rethinking Mau in Colonial Kenya*. New York: Palgrave Macmillan.

Alexander, M. Jacqui. 2005. *Pedagogies of Crossing: Meditations on Feminism, Sexual Politics, Memory, and the Sacred*. Durham, NC: Duke University Press.

Allard, Simplice. 2011. "Libération provisoire de Gbagbo: Deux chefs d'Etat prêts à payer la caution." *Diaspora Côte d'Ivoire*, December 24. https://www.ivoirediaspo.net /liberation-provisoire-de-gbagbo-deux-chefs-detat-prets-a-payer-la-caution/6785 .html.

Allman, Jean. 2004. "Let Your Fashion Be in Line with Our Ghanaian Costume." In *Fashioning Africa: Power and the Politics of Dress*, edited by Jean Allman, 144–65. Bloomington: Indiana University Press.

Amadiume, Ifi. 1987. *Male Daughters and Female Husbands: Gender and Sex in an African Society*. Chicago: University of Chicago Press.

Amadiume, Ifi. 2003. "Prophecy, Authenticity, Oppositional Models: Writers and Politics in Africa." *CR: The New Centennial Review* 3, no. 3 (October): 5–25.

Amadiume, Ifi. 2008. "African Women's Body Images in Postcolonial Discourse and Resistance to Neo-Crusaders." In *Black Womanhood: Images, Icons, and Ideologies of the African Body*, edited by Barbara Thompson and Ifi Amadiume, 49–69. Dartmouth, NH: Hood Museum of Art.

Amon d'Aby, François-Joseph. 1956. *Kwao Adjoba, ou, Procès du régime matriarcal en Basse Côte d'Ivoire, drame en 3 actes et 8 tableaux, suivi d'une note sur la primauté des droits du père*. [*Abidjan, Maison des Combattants, 17 juin 1954*.] Paris: Les Paragraphes littéraires de Paris (impr. de J. Millas-Martin).

Amondji, Marcel. 1984. *Félix Houphouët et la Côte-d'Ivoire: L'envers d'une légende*. Paris: Karthala.

Anderson, Amanda. 2000. "The Temptations of Aggrandized Agency: Feminist Histories and the Horizon of Modernity." *Victorian Studies* 43, no. 1 (October): 43–65.

Anderson, Perry. 1980. *Arguments within English Marxism*. London: NLB.

Andrade, Susan. 2002. "Gender and 'the Public Sphere' in Africa: Writing Women and Rioting Women." *Agenda* 17, no. 54 (January): 45–59.

Appadurai, Arjun. 1986. *The Social Life of Things: Commodities in Cultural Perspective*. Cambridge: Cambridge University Press.

Appadurai, Arjun. 2000. "Grassroots Globalization and the Research Imagination." *Public Culture* 12, no. 1: 1–19

Aravamudan, Srinivas. 2009. "Introduction: Perpetual War." *PMLA* 124, no. 5 (October): 1505–14.

Aravamudan, Srinivas, and Diana Taylor. 2009. "*PMLA*: War." *PMLA* 124, no. 5 (October): 1505–1895.

Ardener, Shirley. 1973. "Sexual Insult and Female Militancy." *Man* 8, no. 3 (September): 422–40.

Arendt, Hannah. [1948] 1973. *The Origins of Totalitarianism*. Boston: Houghton Mifflin Harcourt.

Arendt, Hannah. 1959. *The Human Condition: A Study of the Central Dilemmas Facing Modern Man*. New York: Doubleday.

Arnfred, Signe. 2002. "Simone De Beauvoir in Africa: 'Woman = The Second Sex?' Issues of African Feminist Thought." *Jenda: A Journal of Culture and African Women Studies* 2, no. 1: 1–20.

Arnfred, Signe, ed. 2004. *Re-thinking Sexualities in Africa*. Uppsala, Sweden: Nordic Africa Institute.

Asad, Talal. 1993. *Genealogies of Religion: Discipline and Reasons of Power in Christianity and Islam*. Baltimore: Johns Hopkins University Press.

Asad, Talal. 2000. "Agency and Pain: An Exploration." *Culture and Religion* 1, no. 1 (May): 29–60.

Assié-Lumumba, N'Dri Thérèse. 1996. *Les Africaines dans la politique: Femmes Baoulé de Côte d'Ivoire*. Paris: L'Harmattan.

Associated Press. 2014. "Central African Republic Women Stage Topless Protest Sectarian Violence." *Fox News*, November 22.

Awasom, Susanna Yene. 2003. "Through the Prism of Octogenarian Political Activism in Cameroon: A Critical Survey of the Adoption of Traditional Female Political Institutions to the Exigencies of Modern Politics in Cameroon." In *Indigenous Structures and Governance in Africa*, edited by Olufemi Vaughan, 402–15. Lagos: Sefer Press.

Awoonor, Kofi. 1972. *This Earth, My Brother*. Portsmouth, NH: Heinemann.

Bâ, Mariama. 1981. *Une si longue lettre*. London: Heinemann.

Babatunde, E. D. 2000. "Bini and Yoruba Notions of the Human Personality." In *Substance of African Philosophy*, edited by C. S. Momoh, 274–310. Auchi: African Philosophy Projects' Publications.

Bailleul, Père Charles. 2005. *Sagesse Bambara: Proverbes et sentences*. Bamako: Donniya.

Baker, William Joseph, and James Anthony Mangan, eds. 1987. *Sport in Africa: Essays in Social History*. New York: Africana.

Bal, Mieke. 1996. *Double Exposures: The Subject of Cultural Analysis*. New York: Routledge.

Balibar, Étienne. 2002. *Politics and the Other Scene*. New York: Verso.

Balibar, Étienne. 2015. *Violence and Civility: On the Limits of Political Philosophy*. Translated by G. M. Goshgarian. New York: Columbia University Press.

Banégas, Richard, and Bruno Losch. 2002. "La Côte d'Ivoire au bord de l'implosion." *Politique africaine* 3, no. 87: 139–61.

Barber, Karin, ed. 1997. *Readings in African Popular Culture*. Bloomington: Indiana University Press.

Barcan, Ruth. 2004. *Nudity: A Cultural Anatomy*. Oxford: Berg.

Barlet, Olivier. 2000. *African Cinemas: Decolonizing the Gaze*. London: Zed Books.

Barlet, Olivier. 2005. "Être à la fois africain et contemporain: Entretien avec Jean-Pierre Bekolo." *Africultures*, July. http://www.africultures.com/php/index.php?nav=article&no=3944.

Barlet, Olivier. 2007. "*Les Saignantes, The Bloodletters* by Jean Pierre Bekolo." *Africultures*, June. http://www.africultures.com/index.asp?menu=revue_affiche_article&no=6642&lang=_en.

Barnes, Teresa. 2002. "Owning What We Know: Racial Controversies in South African Feminism, 1991–1998." In *Stepping Forward: Black Women in Africa and the Ameri-*

cas, edited by Catherine Higgs, Barbara A. Moss, and Earline Rae Ferguson, 245–56. Athens: Ohio University Press.

Basden, G. T. [1921] 1966. *Among the Ibos of Nigeria: An Account of the Curious and Interesting Habits, Customs and Beliefs of a Little Known African People by One Who Has for Many Years Lived amongst Them on Close and Intimate Terms.* Reprint. London: Frank Cass.

Basow, Susan A., and Amie C. Braman. 1998. "Women and Body Hair: Social Perceptions and Attitudes." *Psychology of Women Quarterly* 22, no. 4 (December): 637–45.

Basterra, Gabriela. 2004. *Seductions of Fate: Tragic Subjectivity, Ethics, Politics.* Dordrecht: Springer.

Bastian, Misty. 2001. "Dancing Women and Colonial Men: The *Nwaobiala* of 1925." In *"Wicked" Women and the Reconfiguration of Gender in Africa*, edited by Dorothy L. Hodgson and Sheryl A. McCurdy, 109–29. Portsmouth, NH: Heinemann.

Bastian, Misty. 2002. "'Vultures of the Marketplace': Southeastern Nigerian Women and Discourses of the *Ogu Umunwaanyi* (Women's War) of 1929." In *Women in African Colonial Histories*, edited by Jean Allman, Susan Geiger, and Nakanyiki Musisi, 260–81. Bloomington: Indiana University Press.

Bastian, Misty. 2005. "The Naked and the Nude: Historically Multiple Meanings of Oto (Undress) in Southeastern Nigeria." In *Dirt, Undress, and Difference: Critical Perspectives on the Body's Surface*, edited by Adeline Masquelier, 34–60. Bloomington: Indiana University Press.

Bastide, Roger. 1978. *The African Religions of Brazil: Toward a Sociology of the Interpenetration of Civilizations.* Translated by Helen Sebba. Baltimore: Johns Hopkins University Press.

Bax, Pauline. 2013. "Ivory Coast's Women Reject Equality in Household Debate." *Bloomberg*, February 27.

Bax, Pauline, and David Smith. 2011. "Ivory Coast on Brink of Civil War as Seven Women Killed at Protest March." *Guardian*, March 3.

Bayart, Jean-François. 1989. *L'État en Afrique: La politique du ventre.* Paris: Fayard.

BBC News. 2002. "'Deal Reached' in Nigeria Oil Protest." July 16. http://news.bbc .co.uk/2/hi/africa/2129281.stm.

BBC News. 2008. "Ghana to Expel Female Protesters." March 18. http://news.bbc.co.uk /go/pr/fr/-/2/hi/africa/7302243.stm.

BBC News. 2012. "Ivorian Minister Sacked over Toxic Waste Fund Scandal." May 23. https://www.bbc.com/news/world-africa-18173363.

BBC News. 2017. "Stella Nyanzi, the Ugandan Accused of Insulting the President." April 11. https://www.bbc.com/news/world-africa-39558007.

Bekolo, Jean-Pierre, dir. 1992. *Quartier Mozart.* San Francisco: California Newsreel. VHS.

Bekolo, Jean-Pierre, dir. 2005. *Les Saignantes.* Paris: Quartier Mozart Films. DVD.

Bekolo, Jean-Pierre. 2006. "New York Premiere of Jean-Pierre Bekolo's *Les Saignantes.*" *Blogspot*, April 30. http://bekolo.blogspot.com/2006/04/new-york-premiere-of-jean -pierre.html.

Belluck, Pam. 2013. "Comment Ban Sets Off Debate." *New York Times*, September 30.

Benhabib, Seyla. 2002. *The Claims of Culture: Equality and Diversity in the Global Era.* Princeton, NJ: Princeton University Press.

Bennett, Jane. 2009. *Vibrant Matter: A Political Ecology of Things.* Durham, NC: Duke University Press.

Bennett, Michael, and Vanessa D. Dickerson. 2001. "Introduction." In *Recovering the Black Female Body: Self-Representations by African American Women,* edited by Michael Bennett and Vanessa D. Dickerson, 1–18. New Brunswick, NJ: Rutgers University Press.

Berger, John. 1972. *Ways of Seeing.* London: British Broadcasting Corporation and Penguin Books.

Bernault, Florence. 2005. "Body, Power and Sacrifice in Equatorial Africa." *Journal of African History* 47, no. 2 (July): 207–39.

Bester, Rory. 2003. "Floating Free." In *Float: Berni Searle,* 7–55. Cape Town: Bell-Roberts.

Beti, Mongo. 1974. *Remember Ruben.* Paris: Union générale d'éditions.

Beyala, Calixthe. 1988. *Tu t'appelleras Tanga.* Paris: Stock.

Bigley, James, II. 2016. "What It Feels like to Pose Nude for Spencer Tunick." *Cleveland Magazine,* July 18. https://clevelandmagazine.com/in-the-cle/politics/articles/what-it-feels-like-to-pose-nude-for-spencer-tunick.

Birungi, Sandra Janet. 2015. "Stripping Won't Protect Your Land, Minister Tells Women." *Daily Monitor,* August 23. http://www.monitor.co.ug/News/National/Stripping-wont-protect-your-land-minister-tells-women/688334-2842614-7dhhnm/index.html.

Bissell, Mary, dir. 2000. *Naked.* Vancouver: CBC Rough Cuts.

Blier, Suzanne Preston. 1995. *African Vodun: Art, Psychology, and Power.* Chicago: University of Chicago Press.

Bochet de Thé, Marie-Paule. 1985. "Rites et associations traditionnelles chez les femmes béti du sud du Cameroun." In *Femmes du Cameroun: Mères pacifiques, femmes rebelles,* edited by Jean Claude Barbier, 245–83. Paris: Karthala.

Boddy, Janice Patricia. 1989. *Wombs and Alien Spirits: Women, Men, and the Zar Cult in Northern Sudan.* Madison: University of Wisconsin Press.

Bohand, Xavier, C. Monpeurt, S. Bohand, and Alain Cazoulat. 2007. "Toxic Waste and Health Effects in Abidjan City, Ivory Coast." *Médecine Tropicale: Revue du Corps de Santé Colonial* 67, no. 6: 620–24.

Boka, Anicet. 2013. *Coupé-décalé: Le sens d'un genre musical en Afrique.* Paris: L'Harmattan.

Boseley, Sarah. 2008. "Mbeki AIDS Denial 'Caused 300,000 Deaths.'" *The Guardian,* November 26.

Bourgi, Albert, and Christian Casteran. 1991. *Le printemps de l'Afrique: Suivi d'un document du ministère français des affaires étrangères; Scénarios de crise en Afrique.* Paris: Hachette.

Boyce Davies, Carole. 1994. *Black Women, Writing, and Identity: Migrations of the Subject.* New York: Routledge.

Braidotti, Rosi. 2016. "Posthuman Affirmative Politics." In *Resisting Biopolitics: Philosophical, Political, and Performative Strategies*, edited by S. E. Wilmer and Audrone Žukauskaite, 42–68. New York: Routledge.

Brett-Smith, Sarah C. 1995. *The Making of Bamana Sculpture: Creativity and Gender.* Cambridge: Cambridge University Press.

Brodzki, Bella. 2007. *Can These Bones Live? Translation, Survival, and Cultural Memory.* Stanford, CA: Stanford University Press.

Brossard, Dominique, and Dietram A. Scheufele. 2013. "This Story Stinks." *New York Times*, March 2.

Browning, Barbara. 2016. "Why Nakedness Is an Apt Way to Protest the Trump Presidency." *The Conversation*, November 28.

B.s. and G.g. 2002. "Côte d'Ivoire: Coulisses du congrès: L'Adjanou, danse officielle du congrès." *La Voie*, April 10.

Buckley, Thomas, and Alma Gottlieb. 1988. *Blood Magic: The Anthropology of Menstruation.* Berkeley: University of California Press.

Buijtenhuijs, Robert. 1995. "De la sorcellerie comme mode populaire d'action politique." *Politique africaine* 59 (October): 133–39.

Business Insider. 2016. "100 Women Got Naked and Posed across the Street from the Republican National Convention." July 17. https://www.businessinsider.com/afp-100-nude-women-pose-in-cleveland-reflecting-on-trump-2016-7.

Butler, Judith. 1997. *The Psychic Life of Power: Theories in Subjection.* Stanford, CA: Stanford University Press.

Butler, Judith. 2006. *Precarious Life: The Powers of Mourning and Violence.* New York: Verso.

Butler, Judith. 2016. "Rethinking Vulnerability and Resistance." In *Vulnerability in Resistance*, edited by Judith Butler, Zeynep Gambetti, and Leticia Sabsay, 12–27. Durham, NC: Duke University Press.

Byfield, Judith. 2002. *The Bluest Hands: A Social and Economic History of Women Dyers in Abeokuta (Nigeria), 1890–1940.* Portsmouth, NH: Heinemann.

Callimachi, Rukmini. 2011. "Soldiers Open Fire on Women's Protest in Ivory Coast." *Washington Post*, March 3.

Cameron, Vivian. 1991. "Political Exposures: Sexuality and Caricature in the French Revolution." In *Eroticism and the Body Politic*, edited by Lynn Hunt, 90–107. Baltimore: Johns Hopkins University Press.

Cammaert, Jessica. 2016. *Undesirable Practices: Women, Children, and the Politics of the Body in Northern Ghana, 1930–1972.* Lincoln: University of Nebraska Press.

Cantiniaux, David. 2011. "Manifestation pro-Gbagbo à Paris." AFPTV, April 10. https://www.youtube.com/watch?v=-r2IwaPoFHo.

Carlson, Marvin. 1996. *Performance: A Critical Introduction.* New York: Routledge.

Carr-Gomm, Philip. 2010. *A Brief History of Nakedness.* London: Reaktion Books.

Cash, Hilarie, Cosette D. Rae, Ann H. Steel, and Alexander Winkler. 2012. "Internet Addiction: A Brief Summary of Research and Practice." *Current Psychiatry Reviews* 8, no. 4 (November): 292–98.

Césaire, Aimé. 1955. *Discours sur le colonialisme*. Paris: Présence africaine.

Chakrabarty, Dipesh. 1998. "Minority Histories, Subaltern Pasts." *Scrutiny2: Issues in English Studies in Southern Africa* 3, no. 1: 4–15.

Chan, Melissa. 2016. "Topless Women Staged a Protest at Donald Trump's Polling Site." *Fortune*, November 8.

Chatin, Marie-France. 2016. "La CPI, entre faiblesse et lâcheté." RFI, January 22.

Chatterjee, Partha. 1993. *Nation and Its Fragments: Colonial and Postcolonial Histories*. Princeton, NJ: Princeton University Press.

Chausidis, Nikos. 2012. "Mythical Representations of 'Mother Earth' in Pictorial Media." In *An Archaeology of Mother Earth Sites and Sanctuaries through the Ages: Rethinking Symbols and Images, Art and Artefacts from History and Prehistory*, edited by G. Terence Meaden, 5–19. British Archaeological Reports International Series 2389. Oxford: BAR.

Chevilliat, Bernard. 2017. "Le ciel dans le corps: Nudité sacrée et pudeur." *Utreïa! Le corps et le sacré: Nudité, parures, rites, symbolisme . . .* 12 (summer): 81–91.

ChimpReports. 2012. "Photo: Ingrid Breast Attack Protest: 15 Naked Women Arrested." April 23. https://chimpreports.com/4369-photo-ingrid-breast-attack-protest-15-naked-women-arrested/.

Cissé, Souleymane, dir. 1987. *Yeelen=Brightness*. San Francisco, CA: California Newsreel. VHS.

Civox.net. 2012. "Opération 'Kodjo Rouge Terminator' le 20 octobre à Paris." October 14. Accessed July 12, 2016. https://www.civox.net/Operation-Kodjo-Rouge-Terminator-le-20-octobre-a-Paris_a1545.html.

Clark-Bekederemo, John Pepper. 1961. *Song of a Goat*. Ibadan: Mbari.

Cleaver, Tessa, and Marion Wallace. 1990. *Namibia: Women War*. London: Zed Books.

Clinton, Hillary Rodham. 2011. "Secretary Clinton on Violence in Côte d'Ivoire." U.S. Department of State: Diplomacy in Action. March 4. https://2009-2017.state.gov/secretary/20092013clinton/rm/2011/03/157750.htm.

Cobham-Sander, Rhonda. 2003. "Problems of Gender and History in the Teaching of *Things Fall Apart*." In *Chinua Achebe's "Things Fall Apart": A Casebook*, edited by Isidore Okpewho, 16–80. Oxford: Oxford University Press.

Cocks, Tim. 2010. "Newsmaker-Gbagbo Plays Nationalist Card for Ivorian Run-Off." Reuters, November 28. https://af.reuters.com/article/ivoryCoastNews/idAFLDE6AQoB020101128?feedType=RSS&feedName=ivoryCoastNews.

Coetzee, J. M. 1999. *Disgrace*. New York: Penguin.

Cohen, Beth. 1997. "Divesting the Female Breast of Clothes in Classical Sculpture." In *Naked Truth: Women, Sexuality, and Gender in Classical Art and Archaeology*, edited by Ann Olga Koloski-Ostrow, Claire L. Lyons, and Natalie Boymel Kampen, 82–108. New York: Routledge.

Colson, Robert L. 2011. "Arresting Time, Resisting Arrest: Narrative Time and the African Dictator in Ngũgĩ wa Thiong'o's *Wizard of the Crow*." *Research in African Literature* 42, no. 1 (March): 133–53.

Coly, Ayo A. 2010. "A Pedagogy of the Black Female Body: Viewing Angèle Essamba's Black Female Nudes." *Third Text* 24, no. 6 (November): 653–64.

Comaroff, Jean. 1985. *Body of Power, Spirit of Resistance.* Chicago: University of Chicago Press.

Comaroff, Jean. 2007. "Beyond Bare Life: AIDS, (Bio) Politics, and the Neoliberal Order." *Public Culture* 19, no. 1: 197–219.

Comaroff, Jean, and John Comaroff. 1991. *Of Revelation and Revolution: Christianity, Colonialism, and Consciousness in South Africa.* Vol. 1. Chicago: University of Chicago Press.

Conrad, David C. 1999. "Mooning Armies and Mothering Heroes: Female Power in the Manden Epic Tradition." In *Search of Sunjata: The Mande Oral Epic as History, Literature, and Performance,* edited by Ralph A. Austen, 189–230. Bloomington: Indiana University Press.

Conteh-Morgan, John. 1994. *Theatre and Drama in Francophone Africa: A Critical Introduction.* Cambridge: Cambridge University Press.

Cook, Nicolas. 2011. *Côte d'Ivoire Post-Gbagbo: Crisis Recovery.* Washington, DC: Congressional Research Service.

Cook, Rob. 2009. "Liberia Refugees Protest in Ghana." *Pambazuka News,* January 22.

Coombes, Annie. 2001. "Skin Deep/Bodies of Evidence: The Work of Berni Searle." In *Authentic/Ex-Centric: Conceptualism in Contemporary African Art,* edited by Salah M. Hassan and Olu Oguibe, 178–99. Ithaca, NY: Forum for African Arts.

Coombes, Annie E. 2003. *History after Apartheid: Visual Culture and Public Memory in a Democratic South Africa.* Durham, NC: Duke University Press.

Coquery-Vidrovitch, Catherine. 1997. *African Women: A Modern History.* Translated by Beth Gillian Raps. Boulder, CO: Westview.

Corbey, Raymond. 1988. "Alterity: The Colonial Nude." *Critique of Anthropology* 8, no. 3 (December): 75–92.

Cover, Rob. 2003. "The Naked Subject: Nudity, Context and Sexualization in Contemporary Culture." *Body and Society* 9, no. 3 (September): 53–72.

Culler, Jonathan. 1975. *Structuralist Poetics: Structuralism, Linguistics and the Study of Literature.* New York: Routledge.

Dabashi, Hamid. 2012. *The Arab Spring: The End of Postcolonialism.* London: Zed Books.

Dadié, Bernard B. 1950. *Carnet de prison.* Côte d'Ivoire: Centre d'édition et de diffusion africaines (CEDA).

Das, Anwesha. 2014. "A Rendezvous with T. Obinkaram Echewa." *Research in African Literatures* 45, no. 1 (March): 150–60.

Dassé, Claude. 2016. "'Kodjo rouge' pour 'mettre Ouattara dehors' (Henriette Lagou)." *Afrikipresse,* November 6.

Davidow, Bill. 2012. "Exploiting the Neuroscience of Internet Addiction." *The Atlantic,* July 18.

Day, Linda. 2008. "'Bottom Power': Theorizing Feminism and the Women's Movement in Sierra Leone (1981–2007)." *African and Asian Studies* 7, no. 4 (October): 491–513.

de Chalvron, Alain. 2003. "Dominique de Villepin en Côte d'Ivoire." *INA.FR,* January 4. http://www.ina.fr/video/2185400001032.

de la Durantaye, Leland. 2009. *Giorgio Agamben: A Critical Introduction.* Stanford, CA: Stanford University Press.

Deleuze, Gilles, and Félix Guattari. 1987. *A Thousand Plateaus: Capitalism and Schizophrenia*. Translated by Brian Massumi. Minneapolis: University of Minnesota Press.

Denis, Claire, dir. [1999] 2002. *Beau travail*. New York: New Yorker Video. DVD.

Denis, Claire, dir. 2001. *Trouble Every Day*. New York: Kim Stim. DVD.

Denis, Claire, dir. [2004] 2006. *The Intruder*. New York: Fox Lorber Home Video. DVD

Derrida, Jacques. 2002. "The Animal That Therefore I Am (More to Follow)." Translated by David Wills. *Critical Inquiry* 28, no. 2 (January): 369–418.

Dexter, Miriam Robbins, and Starr Goode. 2002. "The Sheela-na-gigs: Sexuality and the Goddess in Ancient Ireland." *Irish Journal of Feminist Studies* 4, no. 2: 50–75.

Dexter, Miriam Robbins, and Victor H. Mair. 2004. "Apotropaia and Fecundity in Eurasian Myth and Iconography: Erotic Female Display Figures." In *Proceedings of the Sixteenth Annual UCLA Indo-European Conference: Los Angeles, November 5–6*, edited by Karlene Jones-Bley, 97–121. Washington, DC: Institute for the Study of Man.

Diabaté, Henriette. 1975. *La marche des femmes sur Grand-Bassam*. Abidjan: Nouvelles Editions Africaines.

Diabate, Naminata. 2016. "Women's Naked Protest in Africa: Comparative Literature and Its Future." In *Theorizing Fieldwork in the Humanities*, edited by Debra Castillo and Shalini Puri, 5–72. Basingstoke, UK: Palgrave Macmillan.

Diabate, Naminata. 2017. "The Cinematic Language of Naked Protest." *Critical Interventions* 11, no. 3 (December): 248–68.

Diabate, Naminata. 2019. "The Forms of Shame and African Literature." In *Routledge Handbook of African Literature*, edited by Moradewun Adejunmobi and Carli Coetzee, 339–53. New York: Routledge.

Diamond, Louise, and John W. McDonald. 1996. *Multi-track Diplomacy: A Systems Approach to Peace*. West Hartford, CT: Kumarian.

Dibba, Lamin. 2001. "Gambia: In Election Ritual Saga Naked Women Protest in Brikama." *The Independent* (Banjul), September 28.

Dibba, Lamin. 2003. "Trente femmes nues et un chien mort dans un drap de satin blanc." *Courrier International*, October 1.

Diduk, Susan. 1989. "Women's Agricultural Production and Political Action in the Cameroon Grassfields." *Africa* 59, no. 3: 338–55.

Diop, Boubacar Boris. 2005. "France in Francophone Africa: Ivory Coast: Colonial Adventure." *Le Monde Diplomatique*, April. https://mondediplo.com/2005/04/10diop.

Disney, Abigail, dir. 2008. *Pray the Devil Back to Hell*. New York: Fork Films. DVD.

Djidjé, M. A. 2003. "Côte d'Ivoire: Nanan Kolia Tanoh, chef du village de Daoukankro (rescapée des danseuses d'Adjanou): 'J'étais sur une liste noire.'" *Fraternité Matin*, March 4. http://fr.allafrica.com/stories/200303040593.html.

Djinko, Nicole Bancouly. 2003. "'Un génocide des Baoulé est mis en oeuvre': Massacre des danseuses d'adjanou L'unique rescapée témoigne." *L'Inter*, March 3.

Donaldson, Thomas. 1975. "Propitious-Apotropaic Eroticism in the Art of Orissa." *Artibus Asiae* 37, no. 1/2: 75–100.

Donkoh, Wilhelmina J. 2015. "Nkrumah and His 'Chicks': An Examination of Women and the Organizational Strategies of the CPP." In *Africa's Many Divides and Africa's Future: Pursuing Nkrumah's Vision of Pan-Africanism in an Era of Globalization*, edited by Charles Quist-Adade and Vincent Dodo, 99–121. Newcastle upon Tyne: Cambridge Scholars.

Doran, D'Arcy. 2002. "Nigeria: Women Claim Victory in ChevronTexaco Oil Terminal Takeover." Associated Press, July 19.

Douglas, Mary. [1966] 2003. *Purity and Danger: An Analysis of the Concepts of Pollution and Taboo*. New York: Routledge.

Doumbia, Yacouba. 2015. "Avant la présidentielle 2015: Banny va lâcher une autre bombe contre le pouvoir." *L'Inter*, March 26.

Dozon, Jean-Pierre. 1985a. "Les Bété: Une création coloniale." In *Au coeur de l'ethnie: Ethnies, tribalisme et état en Afrique*, edited by Jean-Loup Amselle and Elikia M'bolo, 49–85. Paris: La Découverte.

Dozon, Jean-Pierre. 1985b. *La société bété: Histoires d'une ethnie de Côte-d'Ivoire*. Paris: Karthala.

Drewal, Henry John, and Margaret Thompson Drewal. 1983. *Gẹlẹdẹ: Art and Female Power among the Yoruba*. Bloomington: Indiana University Press.

Drewett, Michael, and Martin Cloonan, eds. 2006. *Popular Music Censorship in Africa*. Farnham, UK: Ashgate.

du Plessis, Carien. 2006. "Naked Protest Throws Officials." *News24*, September 14. http://www.news24.com/SouthAfrica/News/Naked-protest-throws-officials -20060914.

Dugger, Celia W. 2008. "Study Cites Toll of AIDS Policy in South Africa." *New York Times*, November 25.

Duval, Philippe. 2003. "Villepin chahuté à Abidjan." *Le Parisien*, January 4.

Echewa, T. Obinkaram. 1976. *The Land's Lord*. Portsmouth, NH: Heinemann.

Echewa, T. Obinkaram. 1986. *The Crippled Dancer*. Portsmouth, NH: Heinemann.

Echewa, T. Obinkaram. [1992] 1993. *I Saw the Sky Catch Fire*. New York: Plume.

Echewa, T. Obinkaram. 1994. *The Ancestor Tree*. London: Dutton Books for Young Readers/Penguin.

Echewa, T. Obinkaram. 1999. *The Magic Tree: A Folktale from Nigeria*. New York: Morrow Junior Books.

The Economist. 1992. "Summers, Lawrence: Let Them Eat Pollution." *The Economist*, February 8, no. 66.

Edgerton, Robert B., and Francis P. Conant. 1964. "Kilipat: The 'Shaming Party' among the Pokot of East Africa." *Southwestern Journal of Anthropology* 20, no. 4: 204–18.

Ehiabhi, Vincent. 2014. "Half Naked Women Protest Gov. Yero's Visit to Sanga." *Legit.ng*. September 23. https://www.legit.ng/295093-half-naked-women-protest-gov-yeros -visit-sanga.html.

Ekine, Sokari. 2008. "The Curse of Nakedness: Women in Nigeria Threaten to Bare It All to Better Their Communities." International Museum of Women. Accessed June 6, 2010. http://exhibitions.globalfundforwomen.org/exhibitions/women-power -and-politics/biology/curse-of-nakedness.

Ekine, Sokari, ed. 2010. *SMS Uprising: Mobile Activism in Africa*. Cape Town: Fahamu/ Pambazuka.

Ellis, Stephen. 1999. *The Mask of Anarchy: The Destruction of Liberia and the Religious Dimension of an African Civil War.* New York: New York University Press.

Ellis, Stephen, and Gerrie ter Haar. 2004. *Worlds of Power: Religious Thought and Political Practice in Africa.* Vol. 1. Oxford: Oxford University Press.

Enwegbara, Basil. 2001. "Toxic Colonialism: Lawrence Summers and Let Africans Eat Pollution." *The Tech* 121, no. 16. http://tech.mit.edu/V121/N16/col16guest.16c.html.

Esposito, Roberto. 2013. *Terms of the Political: Community, Immunity, Biopolitics.* New York: Fordham University Press.

Esposito, Roberto, and Timothy Campbell. 2006. "The Immunization Paradigm." *Diacritics* 36, no. 2: 23–48.

Essuman-Johnson, Abeeku. 2011. "When Refugees Don't Go Home: The Situation of Liberian Refugees in Ghana." *Journal of Immigrant and Refugee Studies* 9, no 2: 105–26.

Estés, Clarissa Pinkola. 1992. *Women Who Run with the Wolves: Myths and Stories of the Wild Woman Archetype.* New York: Ballantine Books.

Etty, Macaire. 2010. "2eme tour de l'élection présidentielle Peuple ivoire . . . L'heure de la grande érection." Abidjan.net, November 22. http://news.abidjan.net/h/380686.html.

Even, André. 1939. "Les propriétés maléfiques et bénéfiques du sexe de la femme selon les croyances des Babamba et Mindassa (Moyen Congo, A.E.F.)." *Bulletin et Memoire de la société d'anthropologie de Paris* 5, 8th series, nos. 1–3: 51–72.

Eventnewstv Presse. 2014. "Exclusif/'Kodjo Rouge' dit de tonnerre le Samedi 7 Juin 2014." YouTube, March 29. https://www.youtube.com/watch?v=PLI0YVZS_GU.

Fabian, Johannes. 2000. *Out of Our Minds: Reason and Madness in the Exploration of Central Africa.* Berkeley: University of California Press.

Fanon, Frantz. [1952] 2008. *Black Skin, White Masks.* New York: Grove.

Fassie, Brenda. 2002. "Vuli Ndlela." *Memeza.* Sandton, South Africa: EMI. CD.

Fellows, Catherine. n.d. "Ivory Coast: Songs and Soldiers." BBC World Service. Accessed November 4, 2018. http://www.bbc.co.uk/worldservice/africa/features/rhythms/ivorycoast.shtml.

Ferguson, James. 2005. "Seeing like an Oil Company: Space, Security, and Global Capital in Neoliberal Africa." *American Anthropologist* 107, no. 3 (September): 377–82.

Ferrari, Federico, and Jean-Luc Nancy. 2014. *Being Nude: The Skin of Images.* New York: Fordham University Press.

Fisiy, Cyprian F. 1998. "Containing Occult Practices: Witchcraft Trials in Cameroon." *African Studies Review* 41, no. 3 (December): 143–63.

Fonchingong, Charles C., Emmanuel Yenshu Vubo, and Maurice Ufon Beseng. 2008. "Traditions of Women's Social Protest Movements and Collective Mobilisation: Lessons from Aghem and Kedjom Women." In *Civil Society and the Search for Development Alternatives in Cameroon,* edited by Emmanuel Yenshu Vubo, 125–41. Dakar: CODESRIA.

Fonte, John. 2008. "Global Governance vs. the Liberal Democratic Nation-State: What Is the Best Regime?" Encounter at 10: The Power of Ideas; the 2008 Bradley Symposium, Washington, DC, June 4.

Forna, Aminatta. 2006. "Speaking in Tongues." *Washington Post,* September 10.

Foua, Ernest de Saint Sauveur. 2011. *Échos de la République du Zouglou*. Abidjan: Balafons.

Foucault, Michel. 1973. *The Order of Things*. New York: Vintage Books.

Foucault, Michel. [1978] 1998. *The History of Sexuality*. Vol. 1, *The Will to Knowledge*. Translated by Robert Hurley. New York: Pantheon Books.

Foucault, Michel. 1979. *Discipline and Punish: The Birth of the Prison*. Translated by Alan Sheridan. New York: Vintage Books.

Foucault, Michel. 1980. *Power/Knowledge: Selected Interviews and Other Writings, 1972–1977*. Translated by Colin Gordon et al. New York: Pantheon Books.

Foucault, Michel. 1982. "The Subject and Power." *Critical Inquiry* 8, no. 4 (summer): 777–95.

Foucault, Michel. 1997. "On the Genealogy of Ethics: An Overview of Work in Progress." In *Ethics, Subjectivity and Truth: The Essential Works of Michel Foucault 1954–1984*, edited by Paul Rabinow, 253–80. New York: New Press.

Foucault, Michel. [1997] 2003. *"Society Must Be Defended": Lectures at the Collège de France, 1975–1976*. Translated by David Macey. Edited by François Ewald and Alessandro Fontana. London: Picador.

Foucault, Michel. 2000. "The Political Technology of Individuals." In *Power: Essential Works of Foucault 1954–1984*, edited by James D. Faubion, 403–17. New York: New Press.

Foucault, Michel. [2004] 2009. *Security, Territory, Population: Lectures at the Collège de France, 1977–78*. Translated by Graham Burchell. London: Picador.

Foucher, Vincent. 2006. "Le 'recours culturel' et la résolution des conflits: l'exemple des Usana en Casamance (Sénégal)." In *Après le conflit, la reconciliation*, edited by Sandrine Lefranc, 313–36. Paris: M. Houdiard.

Foucher, Vincent. 2007. "Tradition africaine et résolution des conflits: Un exemple sénégalais." *Politix* 20, no. 80: 59–80.

Freitag, Barbara. 2004. *Sheela-na-gigs: Unravelling an Enigma*. New York: Routledge.

Frère, Marie-Soleil. 2000. *Presse et démocratie en Afrique francophone: Les mots et les maux de la transition au Bénin et au Niger*. Paris: Karthala.

Frindéthié, K. Martial. 2010. *Globalization and the Seduction of Africa's Ruling Class: An Argument for a New Philosophy of Development*. Jefferson, NC: McFarland.

Fullilove, Michael. 2008. "The World Must Adapt to Diasporas." *Financial Times*, February 14. https://www.ft.com/content/2cff21e8-db33-11dc-9fdd-0000779fd2ac.

Fuphe, Dan. 2001. "South Africa: Eight Women Strip Naked in Protest." *The Sowetan* (Johannesburg), July 13.

Gadeau, Coffi. 1975. "Préface." In *La marche des femmes sur Grand-Bassam*, by Henriette Diabaté, 3. Abidjan: Nouvelles Editions Africaines.

Gandonou, Albert. 2002. *Le roman ouest-africain de langue française: Étude de langue et de style*. Paris: Karthala.

Gardner, Benjamin. 2016. *Selling the Serengeti: The Cultural Politics of Safari Tourism*. Athens: University of Georgia Press.

Gbowee, Leymah. 2011. *Mighty Be Our Powers: How Sisterhood, Prayer, and Sex Changed a Nation at War*. New York: Beast Books.

Gerbner, Katharine. 2015. "Theorizing Conversion: Christianity, Colonization, and Consciousness in the Early Modern Atlantic World." *History Compass* 13, no. 3: 134–47.

Geschiere, Peter. 1995. *Sorcellerie et politique en Afrique: La viande des autres*. Paris: Karthala.

Geschiere, Peter. 2013. *Witchcraft, Intimacy, and Trust: Africa in Comparison*. Chicago: University of Chicago Press.

Ghana Legal. 1994. Public Order Act—1994 (Act 491). http://laws.ghanalegal.com/acts /id/199/public-order-act.

Ghana News Agency. 2008. "Government Decides Liberian Refugees Should Return Home—Bartels." April 2. http://ghananewsagency.org/politics/government -decides-liberian-refugees-should-return-home-bartels-4483.

Ghana Web. 2008. "Government Warns Demonstrating Liberian Refugees." March 8. https://www.ghanaweb.com/GhanaHomePage/election2008/Government-warns -demonstrating-Liberian-refugees-140444#.

Giles, Linda L. 1987. "Possession Cults on the Swahili Coast: A Reexamination of Theories of Marginality." *Africa* 57, no. 2: 234–58.

Gilligan, Carol. 1982. *In a Different Voice: Psychological Theory and Women's Development*. Cambridge, MA: Harvard University Press.

Gilman, Sander L. 1985. "Black Bodies, White Bodies: Toward an Iconography of Female Sexuality in Late Nineteenth-Century Art, Medicine, and Literature." *Critical Inquiry* 12, no. 1: 204–42.

Giroux, Henry A. 2006. "Reading Hurricane Katrina: Race, Class, and the Biopolitics of Disposability." *College Literature* 33, no. 3: 171–96.

Gobodo-Madikizela, Pumla. 2004. *A Human Being Died That Night: A South African Woman Confronts the Legacy of Apartheid*. Boston: Houghton Mifflin Harcourt.

Goffman, Erving. [1963] 2008. "Tightness and Looseness." In *Behavior in Public Places*, 211–13. New York: Simon and Schuster.

Goheen, Miriam. 1996. *Men Own the Fields, Women Own the Crops: Gender and Power in the Cameroon Grassfields*. Madison: University of Wisconsin Press.

Government of the Republic of Liberia. 2008. "Statement by the Government of Liberia in the Wake of Alleged Threats of Reprisal against Ghanaians in Liberia." Ministry of Foreign Affairs, April. http://www.mofa.gov.lr/public2/2press.php?news_id=24 &related=7&pg=sp.

Greenblatt, Stephen. [1991] 2008. *Marvelous Possessions: The Wonder of the New World*. Chicago: University of Chicago Press.

Grillo, Laura S. 2013. "Catachresis in Côte d'Ivoire: Female Genital Power in Religious Ritual and Political Resistance." *Religion and Gender* 3, no. 2: 188–206.

Grillo, Laura S. 2018. *An Intimate Rebuke: Female Genital Power in Ritual and Politics in West Africa*. Durham, NC: Duke University Press.

Grossberg, Lawrence. 1993. "Cultural Studies and/in New Worlds." *Critical Studies in Mass Communication* 10, no. 1: 1–22.

Grosz, Elizabeth. 2006. "Naked." In *The Prosthetic Impulse: From a Posthuman Present to a Biocultural Future*, edited by Marquard Smith and Joanne Morra, 187–202. Cambridge, MA: MIT Press.

Grovogui, Siba N. 2015. "Intricate Entanglement: The ICC and the Pursuit of Peace, Reconciliation and Justice in Libya, Guinea, and Mali." *Africa Development* 40, no. 2: 99–122.

Gueboguo, Charles. 2002. "Manifestations et facteurs explicatifs de l'homosexualité à Yaoundé et à Douala." Master's thesis, Université de Yaoundé.

Guerry, Vincent, and Jean-Pierre Chauveau. 1970. *La vie quotidienne dans un village baoulé: Suivi d' essai bibliographique sur la société baoulé.* Abidjan: INADES (Institut africain pour le développement économique et social).

Guest, Edith M. 1937. "Ballyvourney and Its Sheela-na-gig." *Folklore* 48, no. 4: 374–84.

Guha, Ranajit. [1983] 1994. "The Prose of Counter-insurgency." In *Culture/Power/ History: A Reader in Contemporary Social Theory,* edited by Nicholas B. Dirks, Geoff Eley, and Sherry B. Ortner, 336–71. Princeton, NJ: Princeton University Press.

Gumbu, Maada. 2010. "Secret Societies in Sierra Leone." *Patriotic Vanguard,* August 14. http://www.thepatrioticvanguard.com/secret-societies-in-sierra-leone.

Gunning, Isabelle. 1998. "Cutting through the Obfuscation: Female Genital Surgeries in Neoimperial Culture." In *Talking Visions: Multicultural Feminism in a Transnational Age,* edited by Ella Shohat, 203–24. Cambridge, MA: MIT Press.

Guyer, Jane I. 2004. *Marginal Gains: Monetary Transactions in Atlantic Africa.* Chicago: University of Chicago Press.

Hackett, Rosalind. 1996. *Art and Religion in Africa.* London: Cassell PLC.

Hackett, Rosalind. 2013. "Traditional, African, Religious, Freedom?" *The Immanent France: Secularism, Religion, and the Public Sphere.* January 7. https://tif.ssrc.org /2013/01/07/traditional-african-religious-freedom/.

Hackett, Rosalind, Anne Mélice, Steven Van Wolputte, and Katrien Pype. 2014. "Interview: Rosalind Hackett Reflects on Religious Media in Africa." *Social Compass* 61, no. 1: 67–72.

Hairston, Julia L. 2000. "Skirting the Issue: Machiavelli's Caterina Sforza." *Renaissance Quarterly* 53, no. 3: 687–712.

Hall, Stuart. 1996. "The Question of Cultural Identity." In *Modernity: An Introduction to Modern Societies,* edited by Stuart Hall, David Held, Don Hubert, and Kenneth Thompson, 596–623. Malden, MA: Blackwell.

Hamid, Malick. 2016. "Côte d'Ivoire: Les 'microbes' sèment la panique à Yopougon." *Afrik,* April 24. http://www.afrik.com/Côte-d-ivoire-les-microbes-sement-la -panique-a-yopougon.

Hanna, Judith Lynne. 1988. *Dance, Sex, and Gender: Signs of Identity, Dominance, Defiance, and Desire.* Chicago: University of Chicago Press.

Hardt, Michael, and Antonio Negri. 2000. *Empire.* Cambridge, MA: Harvard University Press.

Hardt, Michael, and Antonio Negri. 2004. *Multitude: War and Democracy in the Age of Empire.* New York: Penguin.

Harrow, Kenneth W. 2002. *Less than One and Double: A Feminist Reading of African Women's Writing.* Portsmouth, NH: Heinemann.

Harrow, Kenneth. 2005. "Let Me Tell You about Bekolo's Latest Film, *Les Saignantes,* but First. . . ." Paper presented at the African Literatures Association Annual Conference, Boulder, CO, April 6–9. http://www.msu.edu/~harrow/recent/LesSaignantes ALApaper.htm.

Harry, Gailey A. 1970. *The Road to Aba: A Study of British Administrative Policy in Eastern Nigeria*. New York: New York University Press.

Harvard Magazine. 2001. "Toxic Memo." May 1. https://harvardmagazine.com/2001/05/toxic-memo.html.

Hazoumé, Paul. 1956. *Le pacte de sang au Dahomey*. Paris: Institut d'ethnologie.

Hegel, Georg Wilhelm Friedrich. 1894. "Geographical Basis of World History." In *Lectures on the Philosophy of World History*, translated by J. Sibree, 83–107. London: Bell.

Henry, Eric. 1999. "The Social Significance of Nudity in Early China." *Fashion Theory* 3, no. 4: 475–86.

Heywood, Linda M. 2017. *Njinga of Angola: Africa's Warrior Queen*. Cambridge, MA: Harvard University Press.

Ho, Erica. 2012. "Better Suit Up: Paris Bans Naked Sunbathers from Tanning in Public." *Time*, August 3.

Hoad, Neville. 2005. "Thabo Mbeki's AIDS Blues: The Intellectual, the Archive, and the Pandemic." *Public Culture* 17, no. 1: 101–28.

Hoad, Neville. 2007. *African Intimacies: Race, Homosexuality, and Globalization*. Minneapolis: University of Minnesota Press.

Hodgson, Dorothy L. 2017. *Gender, Justice, and the Problem of Culture: From Customary Law to Human Rights in Tanzania*. Bloomington: Indiana University Press.

Hodgson, Dorothy L., and Sheryl McCurdy. 1996. "Wayward Wives, Misfit Mothers, and Disobedient Daughters: 'Wicked' Women and the Reconfiguration of Gender in Africa." *Canadian Journal of African Studies/La Revue canadienne des études africaines* 30, no. 1: 1–9.

Hodžić, Saida. 2017. *The Twilight of Cutting: African Activism and Life after NGOs*. Oakland: University of California Press.

Hoffman, Barbara G. 1998. "Secrets and Lies: Context, Meaning, and Agency in Mande (Secrets et mensonges: contexte, signification et rôle des acteurs sociaux mandé)." *Cahiers d'études africaines* 38, no. 149: 85–102.

Hofnung, Thomas. 2003. "Côte-d'Ivoire: Un patriote peut cacher une femme: Nombreuses et organisées, elles affirment promouvoir la paix . . ." *Libération*, February 13. http://www.liberation.fr/monde/2003/02/13/Côte-d-ivoire-un-patriote-peut-cacher-une-femme_430756.

Howard, Philip N., and Muzammil M. Hussain. 2013. *Democracy's Fourth Wave? Digital Media and the Arab Spring*. Oxford: Oxford University Press.

Human Rights Watch (HRW). 2011. "Côte d'Ivoire: Crimes against Humanity by Gbagbo Forces." March 15. https://www.hrw.org/news/2011/03/15/cote-divoire-crimes-against-humanity-gbagbo-forces.

Hunt, Lynn. 2002. "The Many Bodies of Marie Antoinette: Political Pornography and the Problem of the Feminine in the French Revolution." In *The French Revolution: Recent Debates and New Controversies*, 291–313. New York: Routledge.

Hunt, Lynn. 2013. *Family Romance of the French Revolution*. New York: Routledge.

Huntington, Samuel P. 1971. "The Change to Change: Modernization, Development, and Politics." *Comparative Politics* 3, no. 3: 283–322.

Hussain, Nasser. 2007. "Beyond Norm and Exception: Guantánamo." *Critical Inquiry* 33, no. 4: 734–53.

Hutchison, Elizabeth D. 2012. "Spirituality, Religion, and Progressive Social Movements: Resources and Motivation for Social Change." *Journal of Religion and Spirituality in Social Work: Social Thought* 31, nos. 1–2: 105–27.

Ibekwe, Chux, and Akanmu G. Adebayo. 2012. "Dreams and Nightmares: Democratization, Elections, and Conflicts in Africa." In *Managing Conflicts in Africa's Democratic Transitions*, edited by Akanmu G. Adebayo, 3–20. Lanham, MD: Lexington Books.

Ifeka-Moller, Catherine. 1975. "Female Militancy and Colonial Revolt: The Women's War of 1929, Eastern Nigeria." In *Perceiving Women*, edited by Shirley Ardener, 127–57. London: Malaby.

The Independent. 2001. "Naked Women's Protest Is Irreligious." *AllAfrica*, October 5.

IOL. 2006a. "Female Prisoners Strip in Protest." September 7, 2006. https://www.iol .co.za/news/south-africa/female-prisoners-strip-in-protest-292767.

IOL. 2006b. "Naked Women in Prison Hunger Strike." September 8. http://www.iol .co.za/news/south-africa/naked-women-in-prison-hunger-strike-1.292946.

Jackson, Cecile. 2012. "Speech, Gender and Power: Beyond Testimony." *Development and Change* 43, no. 5: 999–1023.

Jaeger, Stephan, Elena Viktorovna Baraban, and Adam Muller, eds. 2012. *Fighting Words and Images: Representing War across the Disciplines.* Toronto: University of Toronto Press.

Jaggi, Maya. 2006. "Decolonise the Mind." *Guardian*, September 9.

Janin, Pierre, and Alain Marie. 2003. "Violences ordinaires, violences enracinées, violences matricielles." *Politique africaine* 3, no. 91: 5–12.

Johnson, Erica L., and Patricia Moran. 2013. "Introduction." In *The Female Face of Shame*, edited by Erica L. Johnson and Patricia Moran, 1–22. Bloomington: Indiana University Press.

Johnson-Odim, Cheryl, and Nina Emma Mba. 1997. *For Women and the Nation: Funmilayo Ransome-Kuti of Nigeria.* Urbana: University of Illinois Press.

Joint Chiefs of Staff. 1993. "Pub 3-53." *Doctrine for Joint Psychological Operations* 30. http://edocs.nps.edu/dodpubs/topic/jointpubs/JP3/JP3_53_930730.pdf.

Jones, Peter. 2015. *Track Two Diplomacy in Theory and Practice.* Stanford, CA: Stanford University Press.

Judeh, Rebecca. 2012. "Bruce Onobrakpeya." *Wordpress: Wolf Pack Introduces Feminism in the Black Diaspora.* November 8. https://rebeccajudeh.wordpress.com/2012/11/08 /bruce-onobrakpeya/.

K. A. 2011. "Yamoussoukro/Usurpation de pouvoir: La révolte des femmes de la cité et des villages." *Le Mandat*, February 17.

Kagan, Robert. 2009. *The Return of History and the End of Dreams.* New York: Vintage.

Kah, Henry Kam. 2011. "Women's Resistance in Cameroon's Western Grassfields: The Power of Symbols, Organization, and Leadership, 1957–1961." *African Studies Quarterly* 12, no. 3: 67–91.

Kalogridis, Jeanne. 2010. *The Scarlet Contessa: A Novel of the Italian Renaissance Novel*. New York: St. Martin's.

Kanogo, Tabitha. 1987. "Kikuyu Women and the Politics of Protest: Mau Mau." In *Images of Women in Peace and War*, edited by Sharon MacDonald, Pat Holden, and Shirley Ardener, 170–89. London: Macmillan Education.

Al-Kassim, Dina. 2007. "Resistance, Terminable and Interminable." In *Derrida, Deleuze, Psychoanalysis*, edited by Gabriele Schwab, 105–41. New York: Columbia University Press.

Keane, Webb. 1997. "From Fetishism to Sincerity: On Agency, the Speaking Subject, and Their Historicity in the Context of Religious Conversion." *Comparative Studies in Society and History* 39, no. 4: 674–93.

Keïta, Cheick Mahamadou Chérif. 2011. *Outcast to Ambassador: The Musical Odyssey of Salif Keita*. Saint Paul, MN: Mogoya Books.

Kelly, James P., III. 2007. "Democratic Evolution and the Church of the United Nations." *Journal of the Federalist Society's Practice Groups* 8, no. 3: 108–13.

Kemedjio, Cilas. 1999. "La modernité africaine et l'écriture de la honte: Le spectre du déshonneur dans les impasses postcoloniales." *L'Esprit créateur* 39, no. 4: 128–38.

Kilgour, Maggie. 1990. *From Communion to Cannibalism: An Anatomy of Metaphors of Incorporation*. Princeton, NJ: Princeton University Press.

King, Helen. 1986. "Agnodike and the Profession of Medicine." *Cambridge Classical Journal* 32: 53–77.

Kirk-Greene, Anthony H. M. 1974. *Mutumin Kirki: The Concept of the Good Man in Hausa*. Bloomington: African Studies Program, Indiana University.

Konan, André Silver. 2012. "Côte d'Ivoire: Le scandale des déchets toxiques emporte le ministre Adama Bictogo." *Jeune Afrique*, May 22.

Konan, Venance. 2006. *Nègreries, 1994–2006: Chroniques de 12 années sèches*. Abidjan: Fraternité Matin.

Konaré, Adame Ba. 2000. *L'os de la parole: Cosmologie du Pouvoir: Essai*. Paris: Présence Africaine.

Konde, Emmanuel. 1990. *The Use of Women for the Empowerment of Men in African Nationalist Politics: The 1958 "Anlu" in Cameroon*. Boston: African Studies Center, Boston University.

Konde, Emmanuel. 2005. *African Women and Politics: Knowledge, Gender, and Power in Male-Dominated Cameroon*. Lewiston, NY: Edwin Mellen.

Koné, Kassim. 2004. "When Male Becomes Female and Female Becomes Male in Mande." *Wagadu* 1 (spring): 1–10.

Kouassi, Albert Konan. 2012. "Cérémonie de purification et de deuil des violences par le CDVR à Bouaké/Charles Konan Banny: 'Que tous les fils de la Côte d'Ivoire renouent la voie du développement de la paix et du dialogue.'" *Ivoire Presse*, August 31.

Kouassi, Charles. 2015. "Présidentielle 2015: Forces et faiblesses de ceux qui veulent ou peuvent être directeurs de campagne de Ouattara." *L'intelligent d'Abidjan*, April 17.

Kouassi, Germain. 2007. *Le phénomène de l'appropriation linguistique et esthétique en littérature africaine de langue française: Le cas des écrivains ivoiriens: Dadié, Kourouma et Adiaffi*. Saint Denis: Publibook.

Kourouma, Ahmadou. 1968. *Les soleils des indépendances*. Montréal: Presses de l'Université de Montréal.

Kourouma, Ahmadou. [1981]. *The Suns of Independence*. Translated by Adrian Adams. London: Heinemann.

Krupat, Arnold. 2002. *Red Matters: Native American Studies*. Philadelphia: University of Pennsylvania Press.

Laburthe-Tolra, Philippe. 1985. *Initiations et sociétés secrètes au Cameroun: Essai sur la religion beti*. Paris: Karthala.

Laburthe-Tolra, Philippe. 1986. *Le tombeau du soleil: Chronique des Bendzo*. Paris: Jacob.

Lafage, Suzanne. 2003. *Le lexique français de Côte-d'Ivoire: Appropriation et créativité*. Le Français en Afrique 17. Nice: Institut de linguistique française/CNRS.

la Hause, Paul. 1982. "Drinking in a Cage: The Durban System and the 1929 Beer Hall Riots." *Africa Perspective* 20: 63–75.

Lamy, Benoit, and Ngangura Mweze, dirs. 1987. *La vie est belle*. Paris: Polygram, Division Polydor. VHS.

Landes, Joan B. 1988. *Women and the Public Sphere in the Age of the French Revolution*. Ithaca, NY: Cornell University Press.

Landes, Joan B. 2003. *Visualizing the Nation: Gender, Representation, and Revolution in Eighteenth-Century France*. Ithaca, NY: Cornell University Press.

Laqueur, Thomas Walter. 1990. *Making Sex: Body and Gender from the Greeks to Freud*. Cambridge, MA: Harvard University Press.

La Tigresa. 2000. "Striptease for the Trees." Westport and Albion, CA, October 18. http://www.earthfilms.org/pages/La%20tigresa/strippress2.html.

Lawrance, Benjamin N. 2003. "La Révolte des Femmes: Economic Upheaval and the Gender of Political Authority in Lomé, Togo, 1931–33." *African Studies Review* 46, no. 1: 43–67.

Lawuyi, Olatunde Bayo. 2012. *Ijapa and Igbin: A Discursive Meditation on Politics, Public Culture and Moral Imagings in Nigeria*. Ibadan: University Press PLC.

Lazzarato, Maurizio. 2002. "From Biopower to Biopolitics." *Pli: The Warwick Journal of Philosophy* 13, no. 8: 1–6.

Le Anh, Tuan. 2004. "An Investigation into the Changes in the Meaning of the Padil Yaya Symbol in the Katu Culture (Chake hamlet Thuong Long-Nam Dong-Thua Thien Hue-Vietnam)." *Vietnam Social Sciences Review* 6, no. 104: 45–60.

Legifrance.gouv.fr. 2016. "Article 222-32 du code pénal." November 17, 2016. Accessed June 18, 2018. https://www.legifrance.gouv.fr/affichCodeArticle.do?idArticle=LEGIARTI000021796944&cidTexte=LEGITEXT000006070719&dateTexte=20100210.

Lemke, Thomas. 2011. *Biopolitics: An Advanced Introduction*. Translated by Eric Frederick Trump. New York: New York University Press.

Leonard, Lori. 2009. "Experiments with 'Modernism': The Allure and the Dangers of Genital Surgeries in Southern Chad." *Medische Antropologie* 21, no. 1: 93–106.

La Lettre du Continent. 2003. "Les amazones de 'Tatie Simone.'" January 30. http://www.africaintelligence.fr/LC-/pouvoirs-et-reseaux/politique/2003/01/30/les-amazones-de-tatie-simone,6069879-ART.

Lévi-Strauss, Claude. [1943] 1963. "The Effectiveness of Symbols." In *Structural Anthropology*, translated by Claire Jacobson and Brooke Grundfest Schoepf, 186–205. New York: Basic Books.

Lichtenstein, Mitchell. 2008. *Teeth*. New York: Weinstein Co., DVD.

Lico, Beniamina. 2014. "Representing the 'Tradition' in Casamance (Senegal): Patterns of Accountability and Ideas of 'Tradition' in the Experience of a Local Development Broker." In *Working the System in Sub-Saharan Africa: Global Values, National Citizenship and Local Politics in Historical Perspective*, edited by Corrado Tornimbeni, 123–41. Newcastle upon Tyne: Cambridge Scholars.

Ligaga, Dina. 2014. "Mapping Emerging Constructions of Good Time Girls in Kenyan Popular Media." *Journal of African Cultural Studies* 26, no. 3: 249–61.

Ligaga, Dina. 2016. "Presence, Agency and Popularity: Kenyan 'Socialites,' Femininities and Digital Media." *East African Literary and Cultural Studies* 2, no. 3: 111–23.

Lorde, Audre. 1983. "The Master's Tools Will Never Dismantle the Master's House.'" In *This Bridge Called My Back: Writings by Radical Women of Color*, edited by Cherríe Moraga and Gloria Anzaldúa, 94–101. New York: Kitchen Table, Women of Color Press.

Loukou, Jean Paul. 2011. "Yamoussoukro: Des femmes dansent l'adjanou pour libérer la Côte d'Ivoire." *Le Nouveau Réveil*, February 9.

Lunceford, Brett. 2012. *Naked Politics: Nudity, Political Action, and the Rhetoric of the Body*. Lanham, MD: Lexington Books.

Maathai, Wangari. 2006. *Unbowed*. New York: Alfred A. Knopf.

Madhavan, Sangeetha, and Aisse Diarra. 2001. "The Blood That Links: Menstrual Regulation among the Bamana of Mali." In *Regulating Menstruation: Beliefs, Practices, Interpretations*, edited by Etienne Van de Walle and Elisha P. Renne, 172–86. Chicago: University of Chicago Press.

Mahmood, Saba. 2009. "Religious Reason and Secular Affect: An Incommensurable Divide?" *Critical Inquiry* 35, no. 4: 836–62.

Maingard, Jacqueline, Sheila Meintjes, and Heather Thompson, dirs. 1995. *Uku Hamba 'Ze (To Walk Naked)*. New York: Third World Newsreel. DVD.

Malagardis, Maria. 2012. "Côte-d'Ivoire valse avec les demons." *Liberation*, January 4.

Mamdani, Mahmood. 1996. *Citizen and Subject: Contemporary Africa and the Legacy of Late Colonialism*. Princeton, NJ: Princeton University Press.

Marcovich, Miroslav. 1986. "Demeter, Baubo, Iacchus, and a Redactor." *Vigiliae christianae* 40, no. 3: 294–301.

Martin, Jean-Baptiste, and François Laplantine, eds. 1994. *Le défi magique*. Vol. 1, *Ésotérisme, occultisme, spiritisme*. Lyon: Presses Universitaires de Lyon.

Marx, Karl. [1852] 1937. *The Eighteenth Brumaire of Louis Bonaparte*. Moscow: Progress.

Masolo, D. A. 2010. *Self and Community in a Changing World*. Bloomington: Indiana University Press.

Mathebula, Mpho. 2018. "Naked Body Protest Is Here to Stay." *News24*, January 28.

Matlwa, Kopano. 2016. *Period Pain*. Auckland Park, South Africa: Jacana.

Matory, J. Lorand. 1993. "Government by Seduction: History and the Tropes of 'Mounting' in Oyo-Yoruba Religion." In *Modernity and Its Malcontents: Ritual and Power in Postcolonial Africa*, edited by Jean Comaroff and John L. Comaroff, 58–85. Chicago: University of Chicago Press.

Maurice, K. Kouassi. 2011. "Crimes des Frci, naufrage économique, crise au sein du Rhdp . . . : Le pouvoir sur le point de craquer." *Le Temps*, December 24.

Mba, Nina Emma. 1982. *Nigerian Women Mobilized: Women's Political Activity in Southern Nigeria, 1900–1965*. Berkeley: Institute of International Studies, University of California.

Mbeki, Thabo. 2016. "Mbeki Addresses 'AIDS Denialism' Criticism." Thabo Mbeki Foundation, March 7. www.thabombekifoundation.org.za.

Mbembe, Achille. 1992. "Provisional Notes on the Postcolony." *Africa: Journal of the International African Institute* 62, no. 1: 3–37.

Mbembe, Achille. 2001. *On the Postcolony: Studies on the History of Society and Culture*. Berkeley: University of California Press.

Mbembe, Achille. 2002. "African Modes of Self-Writing." Translated by Steven Rendall. *Public Culture* 14, no. 1: 239–73.

Mbembe, Achille. 2003. "Necropolitics." *Public Culture* 15, no. 1: 11–40.

Mbembe, Achille. 2013. *Critique de la raison nègre*. Paris: La Découverte.

Mbiti, John S. 1969. *African Religions and Philosophies*. London: Heinemann.

McClintock, Anne. 1995. *Race, Gender, and Sexuality in the Colonial Contest*. New York: Routledge.

Meek, Charles Kingsley. 1937. *Law and Authority in a Nigerian Tribe: A Study in Indirect Rule*. New York: Barnes and Noble.

Meintjes, Sheila. 2007. "Naked Women's Protest, July 1990: 'We Won't Fuck for Houses.'" In *Women in South African History: They Remove Boulders and Cross Rivers*, edited by Nomboniso Gasa, 347–69. Cape Town: Human Sciences Research Council Press.

Mgqolozana, Thando. 2009. *A Man Who Is Not a Man*. Pietermaritzburg: University of KwaZulu-Natal Press.

Miles, Margaret Ruth. 1989. *Carnal Knowing: Female Nakedness and Religious Meaning in the Christian West*. Boston: Beacon.

Mills, Charles W. 2007. "White Ignorance." In *Race and Epistemologies of Ignorance*, edited by Shannon Sullivan and Nancy Tuana, 13–38. Albany: State University of New York Press.

Mills, Charles W. 2015. "Global White Ignorance." In *Routledge International Handbook of Ignorance Studies*, edited by Linsey McGoey and Matthias Gross, 217–27. New York: Taylor and Francis.

Minh-ha, Trinh T., dir. 1982. *Reassemblage: From the Firelight to the Screen*. Berkeley, CA: Trinh T. Minh-ha. VHS.

Ministère de l'éducation nationale et de la recherche scientifique. 1985. *Éducation civique and morale: Le PDCI-RDA, genèse, vie et structures, témoignages, trois discours du président, dix pionniers du parti*. Abidjan: Fraternité Hebdo Éditions.

Minkley, Gary. 1996. "'I Shall Die Married to the Beer': Gender, 'Family' and Space in the East London Locations, 1923–1952." *Kronos* 23 (November): 135–57.

Mintho, Jacquelin. 2011. "Yamoussoukro/Pour avoir été filmées dans leur nudité: Les danseuses de 'l'Adjanou' bloquent le Palais présidentiel." *Le Patriote*, February 9.

Mittelman, James H. 2009. "The Salience of Race." *International Studies Perspectives* 10, no. 1: 99–107.

Moatshe, Rapula. 2016. "ANC Women Bare Bottoms in Protest." *Gauteng*, January 19. http://www.iol.co.za/news/south-africa/gauteng/anc-women-bare-bottoms-in -protest-1972684.

Moi, Toril. 2001. *What Is a Woman? And Other Essays*. Oxford: Oxford University Press.

Moye, David. 2016. "Topless Protesters Disrupt Donald Trump's Polling Place (NSFW): The Women Shouted 'Grab Your Balls! Out of Our Polls!'" *Huffington Post*, November 8.

Mudhai, Okoth F., Wisdom Tettey, and Fackson Banda, eds. 2009. *African Media and the Public Digital Sphere*. New York: Palgrave Macmillan.

Mudimbé, V. Y. 1985. "African Gnosis Philosophy and the Order of Knowledge: An Introduction." *African Studies Review* 28, no. 2/3: 149–233.

Mudimbé, V. Y. 1988. *The Invention of Africa: Gnosis, Philosophy, and the Order of Knowledge*. Bloomington: Indiana University Press.

Mugo, Tiffany Kagure. 2016. "#FeesMustFall: The Threat of the Penis and the Gun in South Africa's Revolutionary Spaces." *Okayafrica*, June 15.

Murphy, Yolanda, and Robert F. Murphy. 1974. *Women of the Forest*. New York: Columbia University Press.

Murray, Margaret Alice. 1934. "Female Fertility Figures." *Journal of the Royal Anthropological Institute of Great Britain and Ireland* 64 (January–June): 93–100.

Musisi, Nakanyike B. 2002. "The Politics of Perception or Perception as Politics? Colonial and Missionary Representations of Baganda Women, 1900–1945." In *Women in African Colonial Histories*, edited by Susan Geiger, Jean Marie Allman, and Nakanyike Musisi, 95–111. Bloomington: Indiana University Press.

Mutua, Makau. 2002. *Human Rights: A Political and Cultural Critique*. Philadelphia: University of Pennsylvania Press.

Mviena, Paul. 1970. *Univers culturel et religieux du peuple béti*. Yaoundé: St.-Paul.

Mwantok, Margaret. 2016. "Bruce Onobrakpeya, the Harmattan Workshop in Retrospective Exhibition." *Guardian Nigeria*, October 6.

Naipaul, V. S. 1984. "The Crocodiles of Yamoussoukro." *New Yorker*, May 14, 52–119.

Nancy, Jean-Luc. 1991. *The Inoperative Community*. Edited by Peter Connor. Translated by Peter Connor, Lisa Garbus, Michael Holland, and Siomna Sawhney. Minneapolis: University of Minnesota Press.

Nancy, Jean-Luc. 1993. *The Birth to the Presence*. Stanford, CA: Stanford University Press.

Nead, Lynda. 2002. *The Female Nude: Art, Obscenity and Sexuality*. New York: Routledge.

Newell, Sasha. 2012. *The Modernity Bluff: Crime, Consumption, and Citizenship in Côte d'Ivoire*. Chicago: University of Chicago Press.

Newell, Stephanie, ed. 2002. *Readings in African Popular Fiction*. Bloomington: Indiana University Press.

Newell, Stephanie, and Onookome Okome, eds. 2013. *Popular Culture in Africa: The Episteme of the Everyday*. New York: Routledge.

The New Humanitarian. 2008. "Cessation Clause Invoked over Refugee Demos." March 20. http://www.thenewhumanitarian.org/report/77397/ghana-liberia -cessation-clause-invoked-over-refugee-demos.

Ngũgĩ wa Thiong'o. 2006. *Wizard of the Crow*. New York: Pantheon.

Niang, Sada. 1996. "Orality in the Films of Sembène Ousmane." In *A Call to Action: The Films of Ousmane Sembène*, edited by Sheila Petty, 67–86. Westport, CT: Praeger.

Nicklaus, Olivier, dir. 2007. *La nudité toute nue*. Paris: Canal+.

Njambi, Wairimũ Ngaruiya. 2004. "Dualisms and Female Bodies in Representations of African Female Circumcision: A Feminist Critique." *Feminist Theory* 5, no. 3: 281–303.

Nkwi, Paul. 1985. "Traditional Female Militancy in a Modern Context." In *Femmes du Cameroun: Mères pacifiques, femmes rebelles*, edited by Jean-Claude Barbier, 181–91. Paris: Karthala-Orstom.

Nnaemeka, Obioma, ed. 2005. *Female Circumcision and the Politics of Knowledge: African Women in Imperialist Discourses*. Westport, CT: Praeger.

Notre Voie. 2010. "Mission du FPI en pays Baoulé—Sakassou réclame le désarmement avant les élections." March 30.

Novo, Salvador. 1994. *The War of the Fatties and Other Stories from Aztec History*. Austin: University of Texas Press.

Obama, Barack. 2011. "Statement by President Barack Obama on the Violence in Côte d'Ivoire." *The White House: President Barack Obama*, March 9. https:// obamawhitehouse.archives.gov/the-press-office/2011/03/09/statement-president -barack-obama-violence-cote-divoire.

O'Carroll, Lisa. 2015. "Sierra Leone's Secret FGM Societies Spread Silent Fear and Sleepless Nights." *The Guardian*, August 25.

Ocungi, Julius. 2015. "Nude Protests: The Acholi Signal of Distress." *Daily Monitor*, March 30.

Ogundipe-Leslie, Molara. 1994. *Re-creating Ourselves: African Women and Critical Transformations*. Trenton, NJ: Africa World Press.

Ogunseitan, Oladipo G. B. 2010. *Be Afra*. Rotherstorpe, UK: Paragon.

Ogunyemi, Chikwenye Okonjo. 1996. *Africa Wo/Man Palava: The Nigerian Novel by Women*. Chicago: University of Chicago Press.

Ogunyemi, Chikwenye Okonjo. 2007. *Juju Fission: Women's Alternative Fictions from the Sahara, the Kalahari, and the Oases In-between*. New York: Peter Lang.

Ohadike, Don. 1996. "Igbo Culture and History." In *Things Fall Apart*, by Chinua Achebe, xix–xlix. New York: Anchor.

Okeke-Ihejirika, Philomina Ezeagbor. 2004. *Negotiating Power and Privilege: Igbo Career Women in Contemporary Nigeria*. Vol. 82. Athens: Ohio University Press.

Okello, Christina. 2015. "Angry Women Strip Bare in Ongoing North Ugandan Land Dispute." RFI, April 19.

Okome, Mojubaolu Olufunke. [1999] 2011. "Listening to Africa, Misunderstanding and

Misinterpreting Africa: Reformist Western Feminist Evangelism on African Women." Unpublished manuscript. Accessed January 15, 2011. https://www.researchgate.net /publication/281440961_Listening_to_Africa_Misunderstanding_and_Misinter preting_Africa_Reformist_Western_Feminist_Evangelism_on_African_Women.

Okonjo, Kamene. 1976. "The Dual-Sex Political System in Operation: Igbo Women and Community Politics in Midwestern Nigeria." In *Women in Africa: Studies in Social and Economic Change*, edited by Kamene Okonjo, Nancy J. Hafkin, and Edna G. Bay, 45–58. Stanford, CA: Stanford University Press.

Okpowo, Blessyn, and Sola Adebayo. 2002. "Itsekiri, Ijaw Women Seize Shell, Chevron Delta Facilities." *All Africa,* November 9.

Olender, Maurice. 1985. "Aspects de Baubo: Textes et contexts antiques." *Revue de l'histoire des religions* 202, no. 1: 3–55.

Olupona, Jacob K. 2014. *African Religions: A Very Short Introduction*. Oxford: Oxford University Press.

Omata, Naohiko. 2012. "Repatriation and Integration of Liberian Refugees from Ghana: The Importance of Personal Networks in the Country of Origin." *Journal of Refugee Studies* 26, no. 2: 265–82.

Ombolo, Jean-Pierre. 1990. *Sexe et société en Afrique noire: L'anthropologie sexuelle beti; Essai analytique, critique et comparatif.* Paris: L'Harmattan.

Onobrakpeya, Bruce. 1994. *Nudes and Protest*. Plastography.

Onobrakpeya, Bruce. 2007. *Nude and Protest: Red Line Experiment*. MutualArt. https:// www.mutualart.com/Artwork/NUDE-AND-PROTEST--RED-LINE -EXPERIMENT-/F25407C4041C50FC.

Oro, Jean-Paul. 2012. "Opération Kodjo rouge 3 à Paris/L'activiste pro-Gbagbo Abel Naki inflamme la mobilisation des femmes patriotes: La police l'arrête pour apologie de la violence et du terrorisme." *L'intelligent d'Abidjan*, March 26.

Osemeka, Irene N. 2011. "The Public Sphere, Women and the Casamance Peace Process." *Historia Actual Online* 25 (spring): 57–65. Accessed July 12, 2015. https:// dialnet.unirioja.es/servlet/articulo?codigo=3670759.

Ouattara, Lacina. 2012. "Opération 'Kodjo rouge' à Paris: Comment le FPI se ridiculise." *Le Patriote*, March 27.

Ousmane, Sembène, dir. 1975. *Xala*. New York: New Yorker Films. VHS.

Oyeniyi, Bukola. 2015. *Dress in the Making of African Identity: A Social and Cultural History of the Yoruba People*. New York: Cambria.

Oyěwùmí, Oyèrónké. 1997. *The Invention of Women: Making an African Sense of Western Gender Discourses*. Minneapolis: University of Minnesota Press.

Oyěwùmí, Oyèrónké. 2003. *African Women and Feminism: Reflecting on the Politics of Sisterhood*. Trenton, NJ: Africa World Press.

Pearson, Ann. 1997. "Reclaiming the Sheela-na-gigs: Goddess Imagery in Medieval Sculptures in Ireland." *Canadian Woman Studies* 17, no. 3: 20–24.

Perham, Dame Margery Freda. 1937. *Native Administration in Nigeria*. Oxford: Oxford University Press.

Piccolino, Giulia. 2014. "The Dilemmas of State Consent in United Nations Peace Operations: The Case of the United Nations Operation in Côte d'Ivoire." In *Peacekeeping in Africa: The Evolving Security Architecture, edited by Marco Wyss and Thierry Tardy, 246–64. New York: Routledge.

Plattner, Marc F. 2008. Democracy without Borders? The Global Challenges to Liberal Democracy. Lanham, MD: Rowman and Littlefield.

Pochon, Caroline, and Allan Rothschild, dirs. 2009. La face cachée des fesses. Issy-les-Moulineaux: Arte TV. DVD.

Povinelli, Elizabeth A. 2002. The Cunning of Recognition: Indigenous Alterities and the Making of Australian Multiculturalism. Durham, NC: Duke University Press.

Pratt, Mary Louise. 2009. "Harm's Way: Language and the Contemporary Arts of War." PMLA 124, no. 5 (October): 1515–31.

Prince, Raymond H. 1961. "The Yoruba Image of the Witch." Journal of Mental Science 107, no. 449: 795–805.

Ramberg, Lucinda. 2014. Given to the Goddess: South Indian Devadasis and the Sexuality of Religion. Durham, NC: Duke University Press.

Ranger, Terence. 1984. "The Invention of Tradition in Colonial Africa." In The Invention of Tradition, edited by E. Hobsbawn and T. Ranger, 57–62. Cambridge: Cambridge University Press.

Reffell, Paul. 2002. "Baring Witness—The New Peace Movement." Baring Witness, December. http://www.baringwitness.org/mission.htm.

Reid, Julian. 2013. "Towards an Affirmative Biopolitics: On the Importance of Thinking the Relations between Life and Error Polemologically." In Foucault, Biopolitics, and Governmentality, edited by Jakob Nilsson and Sven-Olov Wallenstein, 91–104. Huddinge: Södertörns Högskola.

Republic of South Africa: Government Communications. 2001/02. "The Land and Its People." In South African Yearbook 2001/02. Accessed April 29, 2019. https://www.gcis.gov.za/sites/default/files/docs/resourcecentre/yearbook/2002/chap1.pdf.

Rhoades, Georgia. 2010. "Decoding the Sheela-na-gig." Feminist Formations 22, no. 2: 167–94.

Ritzenthaler, Robert E. 1960. "Anlu: A Women's Uprising in the British Cameroons." African Studies 19, no. 3: 151–56.

Robert, William. 2013. "Nude, Glorious, Living." Political Theology 14, no. 1: 115–30.

Roeder, Amy. 2009. "The Cost of South Africa's Misguided AIDS Policies." Harvard T. H. Chan School of Public Health Magazine, spring.

Roosevelt, Theodore. 1910. African Game Trails. New York: Charles Scribner's Sons.

Rosberg, Carl G., Jr., and John Nottingham. [1966] 1985. The Myth of "Mau Mau": Nationalism in Kenya. Nairobi: Transafrica Press.

Rostow, Walt Whitman. 1960. The Stages of Economic Growth: A Non-Communist Manifesto. Cambridge: Cambridge University Press.

Rouch, Jean, dir. 1957. Les maîtres fous. Watertown, MA: Documentary Educational Resources. VHS.

Ruddick, Sara. [1989] 1995. Maternal Thinking: Toward a Politics of Peace. Boston: Beacon.

SABC (South African Broadcasting Corporation). 2012. "Women Strip over Water Shortages." August 29. http://www.sabc.co.za/news/a/4b6f22004c86c1889idedbfc55 0001a1/Women-strip-over-water-shortages-20122908.

SABC (South African Broadcasting Corporation) Digital News. 2016. "Situation Intensifies at Wits as Some Students Stage a Nude Protest." YouTube, October 4. https://www.youtube.com/watch?v=3Om85YRc1WU.

Sadler, Katherine D. 2002. "'Trouble Was Brewing': South African Women, Gender Identity, and Beer Hall Protests, 1929 and 1959." PhD diss., University of California, Los Angeles.

Salzani, Carlo. 2012. "The Notion of Life in the Work of Agamben." *CLCweb: Comparative Literature and Culture* 14, no. 1.

Salyff G. (pseud.) 2011. "Commune d'Abobo: Des femmes essuient des tirs des rebelles, une dizaine de morts." *Le Temps*, March 4.

Salzani, Carlo. 2016. "Nudity: Agamben and Life." *Pléyade* 17 (January–June): 45–64.

Sarkozy, Nicolas. 2007. "Le discours de Dakar de Nicolas Sarkozy." *Le Monde*, September 9.

Sarpong, Awo Abena Amoa, Henrietta Emma Sarpong, and De-Valera N. Y. M. Botchway. 2014. "'Bo Me Truo': A Female-Centred Sun Fire Nudity Dance Ritual of Fertility of the Sehwi People of Ghana." *Chronica Mundi* 9, no. 1: 93–128.

Schaffner, Claire. 2009. "'Les Saignantes,' film Africain d'anticipation." *Afrik*, May 19. http://www.afrik.com/article16792.html.

Schechner, Richard. 1973. "Performance and the Social Sciences: Introduction." *TDR: The Drama Review* 17, no. 3: 3–4.

Schermerhorn, Candace, dir. 2010. *The Naked Option: A Last Resort*. [United States]: Candace Schermerhorn Productions, LLC. DVD.

Schmidt, Hanns-Peter. 1987. *Some Women's Rites and Rights in the Veda*. Poona, India: Bhandarkar Oriental Research Institute.

Schumann, Anne. 2010. "Danse philosophique! The Social and Political Dynamics of Zouglou Music in Abidjan, Côte d'Ivoire, 1990–2008." PhD diss., School of Oriental and African Studies, University of London.

Scott, James C. 1985. *Weapons of the Weak: Everyday Forms of Peasant Resistance*. New Haven, CT: Yale University Press.

Seck, Assane. 2005. *Sénégal, émergence d'une démocratie moderne, 1945–2005: Un itinéraire politique*. Paris: Karthala.

Sefara, Makhudu. 2008. "Will ANC Body Parts Fit?" *City Press*, July 26.

Séhoué, Germain. 1997. *Au nom d'Houphouet: Sacrifices au palais présidentiel*. Paris: L'Harmattan.

Sembène, Ousmane. 1996. *Guelwaar*. Paris: Présence Africaine.

Serubiri, Moses, ed. 2017. *Concerning Nuditude*. n.p.: n.p. www.academia.edu/31698134/BULLET_ConCerning-nuditude.

Shakespeare, William. [1611] 1908. *The Tempest*. New York: Thomas Y. Crowell & Co.

Shaman, David. 1996. "India's Pollution Regulatory Structure and Background." World Bank NIPR Research Paper, January 5.

Shanklin, Eugenia. 1990. "Anlu Remembered: The Kom Women's Rebellion of 1958–61." *Dialectical Anthropology* 15, nos. 2–3: 159–81.

Shannon, Tricia. 2003. "Naked Protest against Naked Aggression." IOL, February 4. https://www.iol.co.za/news/south-africa/naked-protest-against-naked -aggression-101076.

Sharma, Arvind. 1987. "Nudity." In *Encyclopedia of Religion*, edited by Mircea Eliade, 7–10. New York: Macmillan.

Sheehan, Donna. 2002a. "The Genesis of Our Peace Action." Baring Witness. http:// www.baringwitness.org/mission.htm.

Sheehan, Donna. 2002b. "Baring Witness—The Vision." Baring Witness. http://www .baringwitness.org/mission.htm.

Shohat, Ella. 1991. "Imaging Terra Incognita: The Disciplinary Gaze of Empire." *Public Culture* 3, no. 2: 41–70.

Shweder, Richard A. 2012. "Relativism and Universalism." In *A Companion to Moral Anthropology*, edited by Didier Fassin, 85–102. Malden, MA: John Wiley.

Sibi, Aliou, and Henri Konan Bédié. 1995. *Le pari de l'an 2000: Henri Konan Bédié, 35 ans au service de la Côte d'Ivoire*. Abidjan: Agence continentale communication.

Simmel, Georg. [1908] 1950. *The Sociology of Georg Simmel*. Translated by K. H. Wolff. Glencoe, IL: Free Press.

Slaughter, Anne-Marie. 2004. *A New World Order*. Princeton, NJ: Princeton University Press.

Smith, Adam. [1759] 1781. *The Theory of Moral Sentiments, Or, An Essay Towards an Analysis of the Principles by which Men Naturally Judge Concerning the Conduct and Character, First of Their Neighbors, and Afterwards of Themselves: To Which Is Added, a Dissertation on the Origin of Languages*. London: W. Strahan, J. and F. Rivington, T. Longman, and T. Cadell.

Smith, David. 2011. "Ivory Coast Women Defiant after Being Targeted by Gbagbo's Guns." *Guardian*, March 11.

Sonawane, V. H. 1988. "Some Remarkable Sculptures of Lajja Gauri from Gujarat." *Lalit Kala* 23: 27–35.

Souweine, Isaac. 2005. "Naked Protest and the Politics of Personalism." In *Bare Acts*, edited by Monica Narula et al., 526–36. Sarai Reader 5. Delhi: Sarai Programme.

Spencer, Paul. 1988. *The Maasai of Matapato: A Study of Rituals of Rebellion*. Manchester, UK: Manchester University Press.

Spencer, Robert. 2012. "Ngũgĩ wa Thiong'o and the African Dictator Novel." *Journal of Commonwealth Literature* 47, no. 2: 145–58.

Spivak, Gayatri Chakravorty. 1981. "'Draupadi' by Mahasveta Devi: Translator's Foreword." *Critical Inquiry* 8, no. 2: 381–402.

Spivak, Gayatri Chakravorty. 1988. "Can the Subaltern Speak?" In *Marxism and the Interpretation of Culture*, edited by Cary Nelson and Lawrence Grossberg, 271–313. Urbana: University of Illinois Press.

Spivak, Gayatri Chakravorty. 1996. "'Woman' as Theatre: United Nations Conference on Women, Beijing 1996." *Radical Philosophy* 75 (January–February): 2–4.

Spivak, Gayatri Chakravorty. 2009. "Speaking for the Humanities." *Occasion: Interdisciplinary Studies in the Humanities* 1, no. 1: 1–10.

Stam, Robert. 1989. *Subversive Pleasures: Bakhtin, Cultural Criticism, and Film.* Baltimore: Johns Hopkins University Press.

Stanley, Brian. 2003. "Conversion to Christianity: The Colonization of the Mind?" *International Review of Mission* 92, no. 366: 315–31.

Stevens, Phillips, Jr. 2006. "Women's Aggressive Use of Genital Power in Africa." *Transcultural Psychiatry* 43, no. 3: 592–99.

Stoller, Paul. 1995. *Embodying Colonial Memories: Spirit Possession, Power, and the Hauka in West Africa.* New York: Routledge.

Sunday Mail (Brisbane). 2003. "750 Women Go Nude in Protest." February 9.

Swaney, James A. 1994. "So What's Wrong with Dumping on Africa?" *Journal of Economic Issues* 28, no. 2: 367–77.

Swift, Jonathan. 1729. *A Modest Proposal: For Preventing the Children of Poor People in Ireland, from Being a Burden on Their Parents or Country, and for Making Them Beneficial to the Publick.* Dublin: S. Harding. Project Gutenberg, July 27, 2018. https://www.gutenberg.org/files/1080/1080-h/1080-h.htm.

Syrotinski, Michael. 2002. *Singular Performances: Reinscribing the Subject in Francophone African Writing.* Charlottesville: University of Virginia Press.

Tabi, Isidore. 1991. *La théologie des rites Beti: Essai d'explication religieuse des rites Beti et ses implications socio-culturelles.* Yaoundé: Ed. St. Paul.

Talle, Aud. 2007. "'Serious Games': Licences and Prohibitions in Maasai Sexual Life." *Africa* 77, no. 3: 351–70.

Tanga, Pius T. 2006. "The Role of Women's Secret Societies in Cameroon's Contemporary Politics: The Case of Takumbeng." *African Journal of Cross-Cultural Psychology and Sport Facilitation* 8:1–17.

Tansi, Sony Labou. 2016. *The Shameful State.* Bloomington: Indiana University Press.

Taylor, Charles. 2007. *A Secular Age.* Cambridge, MA: Harvard University Press.

Tayoro, P. D. 2002. "Voici les bailleurs de fonds des terroristes." *Notre Voie*, October 12.

Tcheuyap, Alexie. 2005. "African Cinema and Representations of (Homo) Sexuality." In *Body, Sexuality, and Gender*, edited by Flora Veit-Wild and Dirk Naguschewski, 143–54. Versions and Subversions in African Literatures 1. Amsterdam: Rodopi.

Tchouaffe, Olivier. 2006. "Homosexuality and the Politics of Sex, Respectability and Power in Postcolonial Cameroon." *Postamble* 2, no. 2: 4–16.

Teixeira, Maria. 2001. *Rituels divinatoires et thérapeutiques chez les Manjak de Guinée-Bissau et du Sénégal.* Paris: L'Harmattan.

This Is Africa. 2015. "Ugandan Minister Tells Ugandan Women to Stop Nude Protests." August 24.

Tilly, Charles. 2008. *Contentious Performances.* Cambridge: Cambridge University Press.

Tobocman, Seth, Leigh Brownhill, and Teresa E. Turner. 2006. "Nakedness and Power." In *The Best American Comics 2006*, edited by Harvey Pekar and Anne Elizabeth Moore, 204–11. Boston: Houghton Mifflin.

Toerien, Merran, and Sue Wilkinson. 2003. "Gender and Body Hair: Constructing the Feminine Woman." *Women's Studies International Forum* 26, no. 4: 333–44.

Togui, Zéka. 2011a. "Manifestation des femmes patriotes à Paris: Les 'Kodjos Rouges'

en action pour la malédiction effective." *La Dépêche d'Abidjan*, July 26. https://www
.ladepechedabidjan.info/MANIFESTATION-DES-FEMMES-PATRIOTES-A
-PARIS-LES-KODJOS-ROUGES-EN-ACTION-POUR-LA-MALEDICTION
-EFFECTIVE-%E3%80%80_a3853.html.

Togui, Zéka. 2011b. "Opération Kodjo Rouge II à Paris: 'Libérez Gbagbo! . . . Aucune
malédiction n'est de trop.'" *Ivoirebusiness*, September 12.

Togui, Zéka. 2012a. "A Paris, samedi dernier: Des centaines d'Ivoiriens manifestent
devant le QG de Sarkozy." *Notre Voie*, April 3.

Togui, Zéka. 2012b. "Opération Kodjo Rouge III à Paris: Pour la fin du règne sar-
kozien." *La Dépêche d'Abidjan*. March 25. https://www.ladepechedabidjan.info
/OPERATION-KODJO-ROUGE-III-A-PARIS-UNE-TRILOGIE-POUR-LA
-FIN-DU-REGNE-SARKOZIEN_a6786.html.

Togui, Zéka. 2012c. "Paris: Toppo Leontine et les femmes patriotes en Kodjo Rouge
Terminator." *Ivoirebusiness*, October 24. https://www.ivoirebusiness.net/articles
/paris-toppo-leontine-et-les-femmes-patriotes-en-kodjo-rouge-terminator.

Togui, Zéka. 2014. "Interview: Topo Léontine annonce la grande opération Kodjo
rouge international du tonnerre: 'Nous devons chasser Alassane Ouattara de la Côte
d'Ivoire.'" CIVOX.NET, June 3. https://www.civox.net/Interview-Topo-Leontine
-annonce-la-grande-operation-Kodjo-rouge-international-du-tonnerre-Nous-devons
-chasser-Alassane_a4959.html.

Tonda, Joseph. 2005. *Le souverain moderne: Le corps du pouvoir en Afrique centrale
(Congo, Gabon)*. Paris: Karthala.

Toril, Moi. 1985. *Sexual/Textual Politics: Feminist Literary Theory*. New York: Methuen.

Touzard, Philippe. 1987a. *Mémorial de la Côte-d'Ivoire*. Vol. 1, *Les fondements de la nation ivoirienne*. Abidjan: Ami Abidjan.

Touzard, Philippe. 1987b. *Mémorial de la Côte-d'Ivoire*. Vol. 2, *La Côte-d'Ivoire coloniale*.
Abidjan: Ami Abidjan.

Touzard, Philippe. 1987c. *Mémorial de la Côte-d'Ivoire*. Vol. 3, *Du nationalisme à la nation*. Abidjan: Ami Abidjan.

Touzard, Philippe. 1987d. *Mémorial de la Côte-d'Ivoire*. Vol. 4, *Grandes figures ivoiriennes*. Abidjan: Ami Abidjan.

Tripp, Aili Mari, Isabel Casimiro, Joy Kwesiga, and Alice Mungwa. 2009. *African Wom-
en's Movements Transforming Political Landscapes*. Cambridge: Cambridge University
Press.

Tsing, Anna. 2004. *Friction: An Ethnography of Global Connection*. Princeton, NJ:
Princeton University Press.

Tunick, Spencer. 2016. *Everything She Says Means Everything*. Art installation, Cleveland, Ohio, July 17. http://spencertunickcleveland.com/.

Turner, Terence S. 2012. "The Social Skin." *HAU: Journal of Ethnographic Theory* 2, no.
2: 486–504.

Turner, Terisa E., and Leigh S. Brownhill. 2004. "Why Women Are at War with Chevron: Nigerian Subsistence Struggles against the International Oil Industry." *Journal of
Asian and African Studies* 39, nos. 1–2: 63–93.

Turrentine, Jeff. 2006. "The Strongman's Weakness." *New York Times*, September 10.

UNHCR (United Nations High Commissioner for Refugees). 2010. "Protocol and Convention Relating to the Status of Refugees." http://www.unhcr.org/en-us/protection /basic/3b66c2aa10/convention-protocol-relating-status-refugees.html.

Union Comunista Internazionalista. 2003. "Intervention de l'armée française en Côte-d'Ivoire: Derrière les arguments humanitaires, une opération pour protéger les intérêts imperialists." *Lutte de Classe* 70 (January–February). https://www .union-communiste.org/it/2003-01/intervention-de-larmee-francaise-en-cote-divoire -derriere-les-arguments-humanitaires-une.

United States Institute of Peace. 2003. "Côte d'Ivoire: Linas-Marcoussis Agreement." February 14. https://www.usip.org/sites/default/files/file/resources/collections /peace_agreements/Côte_divoire_01242003en.pdf.

UN News. 2008. "UN Refugee Agency Urges Ghana to Halt Deportation of Liberians." March 25. https://news.un.org/en/story/2008/03/253622-un-refugee -agency-urges-ghana-halt-deportation-liberians.

Updike, John. 2006. "Extended Performance: Saving the Republic of Aburiria." *New Yorker* 82, no. 23 (July 31): 74–78.

Vallois, Marie-Claire. 1992. "Exotic Femininity and the Rights of Man: Paul et Virginie and Atala, or the Revolution in Stasis." In *Rebel Daughters: Women and the French Revolution*, edited by Sara E. Melzer and Leslie W. Rabine, 178–97. New York: Oxford University Press.

Van Allen, Judith. 1972. "'Sitting on a Man': Colonialism and the Lost Political Institutions of Igbo Women." *Canadian Journal of African Studies/Revue Canadienne des Études Africaines* 6, no. 2: 165–81.

Van Allen, Judith. 1976. "Aba Riots or the Igbo Women's War? Ideology, Stratification, and the Invisibility of Women." In *Women in Africa: Studies in Social and Economic Change*, edited by Nancy Hafkin and Edna Bay, 59–86. Stanford, CA: Stanford University Press.

Varenne, Leslie. 2012. *Abobo-la-guerre: Côte d'Ivoire, terrain de jeu de la France et de l'ONU*. Paris: Mille et une nuits.

Vatter, Miguel E. 2014. *The Republic of the Living: Biopolitics and the Critique of Civil Society*. New York: Fordham University Press.

Verschueren, Bernard, Clément Tapsoba, and Cheik Kolla Maïga. 1992. "Bekolo: 'The Vision of a People Must Not Be Reduced to Reality.'" *Ecrans d'Afrique* 2: 16–18.

Vilakazi, Thabile, and Brent Swails. 2016. "The Faces behind South Africa's Fees Must Fall Movement." CNN, October 11.

Villa-Vicencio, Charles. 2000. "Why Perpetrators Should Not Always Be Prosecuted: Where the International Criminal Court and Truth Commissions Meet." *Emory Law Journal* 49: 205–22.

Vincent, Isabel. 2016. "Melania Trump like You've Never Seen Her Before." *New York Post*, July 30.

Vincent, Jeanne-Françoise. 1976. *Traditions et transition: Entretiens avec des femmes Beti du Sud-Cameroun*. Paris: ORSTOM.

Vincent, Jeanne-Françoise. [1976] 2001. *Femmes beti entre deux mondes: Entretiens dans la forêt du Cameroun*. Paris: Karthala.

Viola, André, Jacqueline Bardolph, and Denise Coussy. 1998. *New Fiction in English from Africa: West, East, and South*. Amsterdam: Rodopi.

Voice of America. 2006. "Ivory Coast Government Panel Releases Toxic Waste Findings." November 23.

Vuarin, Robert. 2000. *Un système africain de protection sociale au temps de la mondialisation ou "Venez m'aider à tuer mon lion. . . ."* Paris: L'Harmattan.

Wainaina, Binyavanga. 2011. *One Day I Will Write about This Place: A Memoir*. Minneapolis, MN: Graywolf.

Walker, Alice, and Pratibha Parmar. 1993. *Warrior Marks*. New York: Women Make Movies. VHS.

Walker, Benjamin. 1977. *Encyclopedia of Esoteric Man*. New York: Routledge and Kegan Paul.

Walker, Cherryl. [1982] 1991. *Women and Resistance in South Africa*. Cape Town: New Africa Books.

Washington, Teresa. 2005. *Our Mothers, Our Powers, Our Texts: Manifestations of Aje in Africana Literature*. Bloomington: Indiana University Press.

Waylen, Georgina. 1996. *Gender in Third World Politics*. Boulder, CO: Lynne Rienner.

Weeks, John H. 1913. *Among Congo Cannibals: Experiences, Impressions, and Adventures during a Thirty Years' Sojourn amongst the Boloki and Other Congo Tribes, with a Description of Their Curious Habits, Customs, Religion, and Laws*. London: Seeley.

Weheliye, Alexander G. 2014. *Habeas Viscus: Racializing Assemblages, Biopolitics, and Black Feminist Theories of the Human*. Durham, NC: Duke University Press.

Welter, Barbara. 1966. "The Cult of True Womanhood: 1820–1860." *American Quarterly* 18, no. 2: 151–74.

White, Luise. 2000. *Speaking with Vampires: Rumor and History in Colonial Africa*. Berkeley: University of California Press.

Wicomb, Zoë. 1998. "Shame and Identity: The Case of the Coloured in South Africa." In *Writing South Africa: Literature, Apartheid, and Democracy, 1970–1995*, edited by Rosemary Jane Jolly and Derek Attridge, 91–107. Cambridge: Cambridge University Press.

Wipper, Audrey. 1982. "Riot and Rebellion among African Women: Three Examples of Women's Political Clout." In *Perspectives on Power*, edited by Jean O'Barr, 56–62. Durham, NC: Duke University Center for International Relations.

Wiredu, Kwasi, and Kwame Gyekye, eds. 1992. *Person and Community: Ghanaian Philosophical Studies*. Washington, DC: Council for Research in Values and Philosophy.

Witzel, Michael. 2005. "Vala and Iwato: The Myth of the Hidden Sun in India, Japan and Beyond." *Electronic Journal of Vedic Studies* 12, no. 1: 1–69.

World Economic Forum. 2017. *The Global Gender Gap Report: 2017*. Geneva: World Economic Forum.

Wouters, Ruud, and Kirsten Van Camp. 2017. "Less than Expected? How Media Cover Demonstration Turnout." *International Journal of Press/Politics* 22, no. 4: 450–70.

Zabus, Chantal. 2007. *Between Rites and Rights: Excision in Women's Experiential Texts and Human Contexts*. Stanford, CA: Stanford University Press.

Zéré de Mahi (pseud.). 2010. "2e tour du scrutin présidentiel: Pourquoi Ouattara va perdre." *Le Temps*, November 10.

Pager numbers followed by *f* or *t* indicate figures or tables respectively.

postcolonial); colonial vs. postcolonial eras

Povinelli, Elizabeth, 83

Pray the Devil Back to Hell (2008), 45, 108, 127

Prince, Raymond H., 159

prisoners, 21*t*, 75–78, 127

proverbs, 1, 12, 97

Public Order Act: Britain, 195; Ghana, 93–94

"punitive delegation," 11

purification rituals, 13, 66, 67, 73, 85; Adjanou, 141–44; Beti, 108, 142; in *Les Saignantes*, 110, 118–20; Usana, 81

Purity and Danger (1966), 216n8

Quartier Mozart (Bekolo), 113, 208n6

Queer Nation, 41

Ranger, Terence, 85

Ransome-Kuti, Fumilayo, 76

rape, 21–22*t*, 145–47

#RapeAtAzania, 145

Rassemblement des Républicains (Rally of the Republicans, RDR), 72, 155*f*, 212n2

reconciliation. *See* National Reconciliation (2012)

red loincloths. *See* kodjos

Reffell, Paul, 193

refugees, 21*t*, 90–94, 205n2

religions: Islamic, 81, 82, 86, 133, 134, 144; Judeo-Christian, 10–11, 54–55, 74, 81, 82, 86, 133, 144, 200n6

reparations, 13, 134, 143–44

repatriation, 90–91

repression, 89–90

Republic of the Congo (Congo-Brazzaville), 5*f*, 22*t*

"rescue industry," 83

"return of the native," 85–88

Rhodes, Cecil, 144

"Rhodes Must Fall," 144–47

Ritzenthaler, Robert, 199n19

Rob, Cover, 200n25

"romance of resistance," the, 189

Rostow, Walt Whitman, 85

rumors, 21*t*, 73, 81, 95, 101

#RUReferenceList, 145

Rwanda, 90, 104, 195, 207n15

Sacoum, Marguerite, 76

sacralization, 31, 35–36

Saignantes, definition, 24, 122. See also *Les Saignantes* (film)

Sarkozy, Nicolas, 149–51, 154, 156, 157, 160, 167, 170–71

Schermerhorn, Candace, 4, 7, 45, 108

Scott, James C., 160

secret societies, 12, 17, 66, 67, 69, 86; Beti, 108; Cameroonian, 137; condemnation of, 127, 137–38

secularization, 128

Séhoué, Germain, 71

self-exposure, 80, 195

(self-)instrumentalization, 65–80

Senegal, 5*f*, 21*t*, 66, 80, 205n14

Senghor, Fr. Diamacoune, 80, 84

Senghor, Léopold Sédar, 81

Setshwetla (South Africa), 18*t*, 61

sex pollution, 216n8

sex strikes, 40, 41

"sextremism," 9

Shakespeare, William, 163

shame: in African literature, 206n5; personal, 46–48, 89; political uses of, 89–90; proverb about, 97; shame dance, 89–100

shaming. *See* genital shaming

"shaming parties," 11

Sheehan, Donna, 2, 191–92

Shell Oil Company, 140–41

"shooting position." *See* dance

Sibo, Marcelline, 75

"sitting on." *See* genital shaming

"social life," 194

social media, 7–8, 11, 25, 127, 128, 145

social scripting, 50–51

"soft pornography," 53

soldiers. *See* massacres

Somalia, 5*f*, 21*t*

songs: "David contre Goliath," 212n1, 215n28; "Vuli Ndlela," 120–21*f*

South Africa: 5*f*, 21*t*, 43; Cape Town, 193

spirit possession, 53–54, 202n8

spirits: in genitalia, 12, 143; as leaders and helpers, 13, 53, 73, 102, 149, 161–62, 174; in social movements, 160, 162

spirits: evil, 67, 142–43, 153, 179, 184; possession by, 202n8

water shortages, 22*t*, 59–60. *See also* drought
Weapons of the Weak: Everyday Forms of Peasant Resistance (1985), 160
Weheliye, Alexander G., 32
Welcome to the Zoo, 212n17
Weltfriedensdienst (W F D), 85
westernization, 38, 97, 205n14
"white": man, 175, 179, 183; Officer, 175, 179, 184; ignorance, 166–67
white clay and clothing, 67–69, 73, 143, 154, 173
Wicomb, Zoë, 53
widows, 9, 21*t*, 202n6
Wiredu, Kwasi, 293n13
witnessing, 16
Wizard of the Crow (2006), 24, 89, 90, 94–97
Women's March on Grand-Bassam prison, 75–76, 77*f*, 78

"Women's mobbing," 11
Women's War (1929), 22*t*, 25, 76, 175–80, 216n5
World Bank, 34, 85
World Economic Forum, 79

Yoruba, 12, 159, 162
Yorubaland, 22*t*, 178
Young Patriots, 68
YouTube, 8, 11, 30, 128, 145, 146, 158, 173

Zabus, Chantal, 16
Zambia, 5*f*, 23*t*
Ziguinchor, 81
Zimbabwe, 5*f*, 23*t*, 87, 98
zoe, 39–40, 116
zombie-likeness, 43, 122
Zulu, 18*t*, 46